Lecture Notes in Artificial Intelligence 9575

Subseries of Lecture Notes in Computer Science

LNAI Series Editors

Randy Goebel
  University of Alberta, Edmonton, Canada
Yuzuru Tanaka
  Hokkaido University, Sapporo, Japan
Wolfgang Wahlster
  DFKI and Saarland University, Saarbrücken, Germany

LNAI Founding Series Editor

Joerg Siekmann
  DFKI and Saarland University, Saarbrücken, Germany

More information about this series at http://www.springer.com/series/1244

Katsumi Inoue · Hayato Ohwada
Akihiro Yamamoto (Eds.)

Inductive Logic Programming

25th International Conference, ILP 2015
Kyoto, Japan, August 20–22, 2015
Revised Selected Papers

 Springer

Editors
Katsumi Inoue
National Institute of Informatics
Tokyo
Japan

Akihiro Yamamoto
Kyoto University
Kyoto
Japan

Hayato Ohwada
Tokyo University of Science
Noda, Chiba
Japan

ISSN 0302-9743   ISSN 1611-3349   (electronic)
Lecture Notes in Artificial Intelligence
ISBN 978-3-319-40565-0   ISBN 978-3-319-40566-7   (eBook)
DOI 10.1007/978-3-319-40566-7

Library of Congress Control Number: 2016941080

LNCS Sublibrary: SL7 – Artificial Intelligence

This Springer imprint is published by Springer Nature
The registered company is Springer International Publishing AG Switzerland

Preface

This volume contains the revised versions of selected papers presented at ILP 2015: The 25th International Conference on Inductive Logic Programming. ILP 2015 was held in Kyoto University, Kyoto, Japan, during August 20–22, 2015. The ILP conference series has been the premier international forum on ILP. Topics in ILP conferences address theories, algorithms, representations and languages, systems and applications of ILP, and cover all areas of learning in logic, relational learning, relational data mining, statistical relational learning, multi-relational data mining, relational reinforcement learning, graph mining, connections with other learning paradigms, among others.

We solicited three kinds of papers: (1) long papers describing original mature work containing appropriate experimental evaluation and/or representing a self-contained theoretical contribution, (2) short papers describing original work in progress, brief accounts of original ideas without conclusive experimental evaluation, and other relevant work of potentially high scientific interest but not yet qualifying for the long paper category, and (3) papers relevant to the conference topics and recently published or accepted for publication by a first-class conference or journal.

There were 44 submissions in total, 24 long papers, 18 short papers, and two published papers. Long papers were reviewed by at least three members of the Program Committee (PC), and 13 papers were accepted for oral presentation at ILP 2015. Among them, five long papers were invited to submit a final version to the LNAI conference proceedings (without a further review) during the first round of reviews, and the other eight papers were invited to submit a full version to the proceedings after another review round. Short papers were firstly reviewed on the grounds of relevance by PC co-chairs, and 17 papers were accepted for short oral presentation. Each short paper was reviewed by at least three members of the PC, and 11 papers were invited to submit a full version to the proceedings after another review round. Among the 19 invited papers that needed second reviews, 16 papers were finally submitted. Each submission was reviewed again by at least three PC members, and nine papers were accepted finally. Hence, together with the five long papers without the second reviews, the PC co-chairs finally decided to include 14 papers in this volume. This is a rather complicated review process with the three review rounds (long, short, and proceedings), but we believe that we can select high-quality papers this way. We thank the members of the PC for providing high-quality and timely reviews.

There were 10 technical sessions at ILP 2015: Nonmonotonic Semantics, Logic and Learning, Complexity, Action Learning, Distribution Semantics, Implementation, Kernel Programming, Data and Knowledge Modeling, and Cognitive Modeling. Two proceedings are published for ILP 2015: a volume of Springer's LNAI series for selected papers (this volume) and an electronic volume of CEUR-WS.org for late-breaking papers. Moreover, there will be a special issue on ILP in the journal *Machine Learning*.

The program of ILP 2015 also included three excellent invited talks given by Stephen Muggleton from Imperial College London, Taisuke Sato from Tokyo Institute of Technology, and Luc De Raedt from Katholieke Universiteit Leuven. Stephen Muggleton gave the talk "Meta-Interpretive Learning: Achievements and Challenges," and detailed their work on meta-interpretive learning, which is a recent ILP technique aimed at supporting learning of recursive definitions and predicate invention. Taisuke Sato first published the distribution semantics for probabilistic logic programming (PLP) in 1995, and ILP 2015 celebrated the 20th anniversary of the distribution semantics in the form of Sato's talk "Distribution Semantics and Cyclic Relational Modeling," which was followed by a session of probabilistic ILP. Luc De Raedt reported in his invited talk "Applications of Probabilistic Logic Programming" their recent progress in applying PLP to challenging applications.

At ILP 2015, the *Machine Learning* journal generously continued its sponsorship of the best student paper award. The two best student paper awards of ILP 2015 were given to Golnoosh Farnadi for her paper "Statistical Relational Learning with Soft Quantifiers" (co-authored with Stephen H. Bach, Marjon Blondeel, Marie-Francine Moens, Martine De Cock, and Lise Getoor), and Francesco Orsini for his paper "kProlog: An Algebraic Prolog for Kernel Programming" (co-authored with Paolo Frasconi and Luc De Raedt).

To celebrate the 25th anniversary of the ILP conference series, ILP 2015 organized a panel discussion on the past and future progress of ILP. The panelists were Stephen Muggleton, Fabrizio Riguzzi, Filip Zelezny, Gerson Zaverucha, Jesse Davis, and Katsumi Inoue, who are all chairs of the last five years of ILP conferences (2011–2015), and Taisuke Sato.

A survey of ILP 2015 including the abstracts of these three invited talks and "ILP 25 Years Panel" as well as recent trends in ILP was presented at AAAI-16 as a "What's Hot" talk—Katsumi Inoue, Hayato Ohwada, and Akihiro Yamamoto, "Inductive Logic Programming: Challenges," in: *Proceedings of the 30th AAAI Conference on Artificial Intelligence* (AAAI-16; Phoenix, Arizona, USA, February 14, 2016), pp. 4330–4332, 2016.

ILP 2015 was kindly sponsored by The Japanese Society for Artificial Intelligence (JSAI), *Artificial Intelligence* (Elsevier), *Machine Learning* (Springer), Support Center for Advanced Telecommunications Technology Research Foundation (SCAT), Inoue Foundation for Science, SONAR Ltd., Video Research Ltd., The Graduate University for Advanced Studies (SOKENDAI), National Institute of Informatics (NII), Tokyo University of Science, and Kyoto University. Last but not least, we would like to thank the members of the Local Committee of ILP 2015: Kotaro Okazaki (local chair), Taku Harada, Kimiko Kato, Hiroyuki Nishiyama, Tony Ribeiro, Suguru Ueda, Ryo Yoshinaka, and their teams. They did an outstanding job with the local arrangements, and the conference would not have been possible without their hard work.

April 2016   Katsumi Inoue
                                                                    Hayato Ohwada
                                                                    Akihiro Yamamoto

Organization

Program Committee

Erick Alphonse	LIPN - UMR CNRS 7030, France
Annalisa Appice	University of Bari Aldo Moro, Italy
Elena Bellodi	University of Ferrara, Italy
Hendrik Blockeel	K.U. Leuven, Belgium
Rui Camacho	LIACC/FEUP University of Porto, Portugal
James Cussens	University of York, UK
Jesse Davis	K.U. Leuven, Belgium
Luc De Raedt	K.U. Leuven, Belgium
Inês Dutra	CRACS INES-TEC LA and Universidade do Porto, Portugal
Saso Dzeroski	Jozef Stefan Institute, Slovenia
Nicola Fanizzi	University of Bari Aldo Moro, Italy
Stefano Ferilli	University of Bari Aldo Moro, Italy
Peter Flach	University of Bristol, UK
Nuno A. Fonseca	EMBL-EBI, European Bioinformatics Institute, UK
Tamas Horvath	University of Bonn and Fraunhofer IAIS, Germany
Katsumi Inoue	NII, Japan
Nobuhiro Inuzuka	Nagoya Institute of Technology, Japan
Andreas Karwath	University of Mainz, Germany
Kristian Kersting	TU Dortmund University, Germany
Ross King	University of Manchester, UK
Ekaterina Komendantskaya	University of Dundee, UK
Nada Lavrač	Jozef Stefan Institute, Slovenia
Francesca Alessandra Lisi	University of Bari Aldo Moro, Italy
Donato Malerba	University of Bari Aldo Moro, Italy
Stephen Muggleton	Imperial College London, UK
Sriraam Natarajan	Indiana University, USA
Hayato Ohwada	Tokyo University of Science, Japan
Aline Paes	Universidade Federal Fluminense, Brazil
Bernhard Pfahringer	University of Waikato, New Zealand
Ganesh Ramakrishnan	IIT Bombay, India
Jan Ramon	K.U. Leuven, Belgium
Oliver Ray	University of Bristol, UK
Fabrizio Riguzzi	University of Ferrara, Italy
Celine Rouveirol	LIPN, Université Paris 13, France
Alessandra Russo	Imperial College London, UK
Chiaki Sakama	Wakayama University, Japan

Vítor Santos Costa	Universidade do Porto, Portugal
Takayoshi Shoudai	Kyushu International University, Japan
Ashwin Srinivasan	BITS-Pilani, India
Alireza Tamaddoni-Nezhad	Imperial College, London, UK
Tomoyuki Uchida	Hiroshima City University, Japan
Guy Van den Broeck	K.U. Leuven, Belgium
Jan Van Haaren	K.U. Leuven, Belgium
Christel Vrain	LIFO - University of Orléans, France
Stefan Wrobel	Fraunhofer IAIS and University of Bonn, Germany
Akihiro Yamamoto	Kyoto University, Japan
Gerson Zaverucha	PESC/COPPE - UFRJ, Brazil
Filip Zelezny	Czech Technical University, Czech Republic

Additional Reviewers

Côrte-Real, Joana
Ribeiro, Tony
Sato, Taisuke
Warburton, Chris

Contents

Relational Kernel-Based Grasping
with Numerical Features

Laura Antanas[⊠], Plinio Moreno, and Luc De Raedt

Department of Computer Science, KU Leuven, Leuven, Belgium
{laura.antanas,plinio.moreno,luc.deraedt}@cs.kuleuven.be

Abstract. Object grasping is a key task in robot manipulation. Performing a grasp largely depends on the object properties and grasp constraints. This paper proposes a new statistical relational learning approach to recognize graspable points in object point clouds. We characterize each point with numerical shape features and represent each cloud as a (hyper-) graph by considering qualitative spatial relations between neighboring points. Further, we use kernels on graphs to exploit extended contextual shape information and compute discriminative features which show improvement upon local shape features. Our work for robot grasping highlights the importance of moving towards integrating relational representations with low-level descriptors for robot vision. We evaluate our relational kernel-based approach on a realistic dataset with 8 objects.

Keywords: Robot grasping · Graph-based representations · Numerical shape features · Relational kernels · Numerical feature pooling

1 Introduction

To operate in the real world, a robot requires good manipulation skills. A good robot grasp depends on the specific manipulation scenario, and essentially on the object properties, as well as grasp constraints (e.g., gripper configuration, environmental restrictions). As in robot manipulation objects are widely described using point clouds, robot grasping often relies on finding good mappings between gripper orientations and object regions (or points). To this end, much of the current work on robot grasping focuses on adapting low-level descriptors popular in the computer vision community (i.e., shape context) to characterize the graspability of an object point. Essentially, this translates into calculating, for each point in the cloud, a shape feature descriptor that summarizes a limited neighbouring surface around the point. However, such local shape features do not work properly on very complex or (self-) occluded objects.

A first contribution of this paper is to investigate whether *the structure of the object can improve robot grasping by means of statistical relational learning (SRL)*. In order to do so, we propose to employ a graph-based representation of the object that exploits both local numerical shape features and higher-level

© Springer International Publishing Switzerland 2016
K. Inoue et al. (Eds.): ILP 2015, LNAI 9575, pp. 1–14, 2016.
DOI: 10.1007/978-3-319-40566-7_1

information about the structure of the object. Given a 3D point cloud of the object, we characterize each point with shape features and represent the cloud as a (hyper-) graph by adding symbolic spatial relations that hold among neighboring object points. As a result, graph nodes corresponding to object points are characterized by distributions of numerical shape features instead of semantic labels. The derived relational graph captures extended contextual shape information of the object which may be useful to better recognize graspable points. As an example, consider a graspable point on the rim of a cup. Although it may be characterized by a misleading local shape descriptor due to its position or perceptual noise, this can be corrected by nearby graspable points with more accurate shape features.

As a second contribution, we propose a *new relational kernel-based approach to numerical feature pooling* for robot grasping. To recognize graspable points we employ relational kernels defined on the attributed graph. For each point, our relational kernel exploits extended contextual information and aggregates (or pools) numerical shape features according to the graph structure, yielding more discriminative features. Its benefit is shown experimentally on a realistic dataset. Our work highlights the importance of moving towards integrating relational representations with low-level descriptors for robot vision.

We proceed as follows. We first explain in Sect. 2 the grasping primitives that define our setup. Afterwards, we present our relational formulation for the learning problem considered (Sect. 3) and show how we solve it with variants of relational kernels (Sect. 4). Next, in Sect. 5 we present our experimental results. Before concluding, we review related work on robot grasping, feature pooling and graph kernels (Sect. 6).

2 The Robot Grasping Scenario and Grasping Primitives

We consider the robot scenario in Fig. 1. The robotic platform is next to a table and on the table there are one or more objects for grasping exploration. The robot has the following components: a mobile component, an arm, a gripper and a range camera. An object (e.g., cup, glass) may be placed on the table at various poses. Each pose provides a point cloud, obtained via the range sensor. The points above the table are converted, using segmentation techniques (e.g., [24]), into a point cloud describing the object. Figure 1 illustrates the point cloud of the visible side of a cup placed on the table sideways. The goal is to determine the pre-grasp pose, that is, where to place the gripper with respect to the object in order to execute a stable grasp. Motion planning from the current gripper pose to the pre-grasp pose reduces the number of grasping hypotheses due to kinematic and environmental constraints. The reduced set of reachable local regions provides the data samples for learning to recognize graspable object points.

We consider three types of domain primitives which we use to build our relational representation (or hyper-graphs) of the grasping problem: *reaching points*, their *3D locations* and their numerical *shape features*. Reaching points are labeled using the simulator. The robot executes grasps on the object points and if they are successful, the reaching points become positive instances. Next, each reaching point is characterized by several local 3D shape features computed in its neighborhood. The neighborhood of each point consists of a 3D grid centred at the reaching point and oriented with respect to the projection of the point's normal on the table plane and the gravity vector, as illustrated in Fig. 1. We consider as neighborhood grid, in turn, a gripper cell and a sphere around the point and calculate three shape features: 3D shape context (SC) [18], point feature histogram (PFH) [27] and viewpoint feature histogram (VFH) [28].

 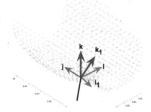

Fig. 1. Robot grasping scenario. The gripper and objects on the table (left). A partial point cloud of a can placed on the table (right). The (i, j, k) is the reference frame of the camera centred at the sample point. Its normal is the black line. The (i_1, j, k_1) is the reference frame of the 3D grid, which is obtained by rotating the (i, j, k) frame along the y axis.

While the PFH feature encodes the statistics of the shape of a point cloud by accumulating the geometric relations between all point pairs, the VFH augments PFH with the relation between the camera's point of view and the point cloud of an object. The 3D SC describes the shape of the object as quantitative descriptions centered at different points on the surface of the object. The shape context of a point is a coarse histogram of the relative coordinates of the remaining surface points. The bins of the histogram are constructed by the overlay of concentric shells around the center point and sectors emerging from this point.

3   Relational Grasping: Problem Formulation

Next, we represent the grasping primitives as a relational database and use it as input to our relational learning system. We use the kLog framework [11] to build our relational kernel-based approach to grasping point recognition. Embedded in Prolog, kLog is a domain specific language for kernel-based learning, that allows to specify in a declarative way relational learning problems. It learns from interpretations [8], transforms the relational databases into graph-based representations and uses graph kernels to extract the feature space.

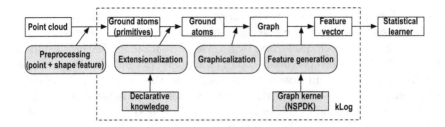

Fig. 2. From point clouds to feature vectors in kLog.

Figure 2 illustrates the information flow in kLog for robot grasping. We model our graspable point recognition problem starting from the grasping primitives which we represent as relational databases. Next, we define declaratively spatial relations between reaching points. The extended relational database is used by kLog to build kernel features which are finally used for learning. We explain in more detail each step for our grasping problem.

3.1 Data Modeling

Grasping primitives are represented at a higher level using a relational language derived from its associated entity/relationship (E/R) data model, as in database theory [12], with some further assumptions required by kLog. It is based on entities, relationships linking entities and attributes that describe entities and relationships. Figure 3(a) shows the E/R diagram for our grasping point problem. A *reaching entity* is any reaching point. It is represented by the relation point$(id, f_1, \ldots, f_n)$, which indicates that it has a unique identifier id (underlined oval) and shape properties. The vector $[f_1, \ldots, f_n]$ represents a shape feature characterizing the reaching point. Each f_i is a shape feature vector component and is represented as an entity attribute. For example, the tuple point$(p_1, 10.8, \ldots, 557.9)$ specifies a specific reaching point entity (depicted as rectangle in Fig. 3(b)), where p_1 is its identifier and the other arguments are shape feature components.

Relationships are qualitative *spatial relations* among entities (diamonds) and are derived from their 3D spatial locations. They impose a structure on reaching entities. In practice, we employ the relationship closeBy2(p_1, p_3) which indicates that reaching entities p_1 and p_3 are spatially close to each other, and the relationship closeBy3(p_1, p_2, p_3) which indicates that reaching entities p_1, p_2 and p_3 are spatially close to each other. A special relationship is introduced by the predicate category$(id, class)$ (dashed diamond). It is linked to reaching entities and associates a binary class label *grasp/nonGrasp* to each entity, indicating if the reaching point is graspable or not.

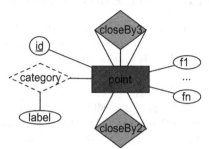

(a) Proposed E/R scheme: rectangles denote entity vertices, diamonds denote relationships, and circles (except point id) denote local properties.

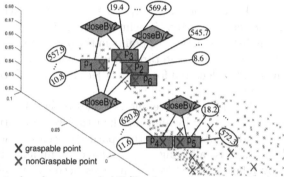

(b) Part of a *glass* grounded E/R scheme mapped on its point cloud.

$x = \{\texttt{point}(p_1, 10.8, \ldots, 557.9), \texttt{point}(p_2, 8.6, \ldots, 545.7), \texttt{point}(p_3, 19.4, \ldots, 569.4),$
$\texttt{point}(p_4, 11.6, \ldots, 620.8), \texttt{point}(p_5, 18.2, \ldots, 572.3), \ldots, \texttt{closeBy2}(p_1, p_3),$
$\texttt{closeBy2}(p_3, p_2), \texttt{closeBy2}(p_4, p_5), \ldots, \texttt{closeBy3}(p_1, p_2, p_3), \ldots \}.$
$y = \{\texttt{category}(p_1, nonGrasp), \texttt{category}(p_2, nonGrasp), \texttt{category}(p_3, nonGrasp),$
$\texttt{category}(p_4, grasp), \texttt{category}(p_5, grasp), \ldots \}.$

(c) Point cloud interpretation $i = (x, y)$ of a *glass* point cloud.

Fig. 3. Relational robot grasping in kLog. (Color figure online)

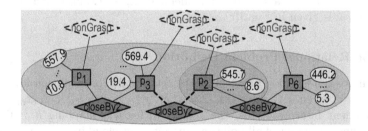

Fig. 4. From point cloud graph to feature vectors in kLog.

3.2 Declarative and Relational Feature Construction

We define the spatial relations using logical rules. For example, the relation
closeBy2/2 holds between two points that belong to the same point cloud and
are spatially close to each other. It can be defined as follows:

$$\text{closeBy2}(P_1, P_2) \leftarrow \text{point}(P_1, F_{11}, \ldots, F_{1n}), \text{point}(P_2, F_{21}, \ldots, F_{2n}),$$
$$\text{cloud}(P_1, V), \text{cloud}(P_2, V), P_1 \leq P_2,$$
$$\text{objectLength}(L), \text{objectHeight}(H), \text{objectWidth}(W),$$
$$T_x = c * L, T_y = c * H, T_z = c * W,$$
$$\text{edist}(P_1, P_2, D_x, D_y, D_z), D_x < T_x, D_y < T_y, D_z < T_z.$$

The condition $\text{cloud}(P_1, V)$, $\text{cloud}(P_2, V)$ specifies that P_1 and P_2 belong to
the same point cloud V. The inequality $P_1 \leq P_2$ removes the symmetry of the
close by relation. The relation $\text{edist}/5$, defined in a similar way, represents the
normalized Euclidian distance between 2 points in the 3D space. As the definition
shows, it is projected on all 3 axes and thresholded on each axis i. The thresholds
T_i are distance thresholds calculated for every object from the object dimensions
using a constant ratio c.

The close by relation defined above allows cycles of size 3 or greater in the
graph. We enforce more sparsity by allowing the closeBy2/2 relation between 2
points to hold if there does not exist another path between the two points that
involves another node, thus, allowing only cycles of minimum size 4. We use
the sparser close by relation in practice as it gives better results. If we denote
the previous closeBy2/2 relation as closeBy2_initial/2, the sparser relation
is defined as:

$$\text{closeBy2}(P_1, P_2) \leftarrow \text{closeBy2\_initial}(P_1, P_2), L = 3, \text{not } \text{path}(P_1, P_2, L).$$

The relation $\text{path}/3$ checks if there is a path smaller than or equal to 2 edges
between nodes P_1 and P_2. We define in a similar way the relation closeBy3/3
which holds between three points that belong to the same point cloud and are
spatially close to each other.

In our setting each point cloud is represented as an instance of a relational
database (i.e., as a set of relations), and thus, as a *point cloud interpretation*.
Object point clouds are assumed to be independent. An example of a point cloud
interpretation is given in Fig. 3(c).

3.3 The Relational Problem Definition

We formulate the learning problem at the relational representation level in the
following way: given a training set $D = \{(x_1, y_1), \ldots, (x_2, y_2), \ldots, (x_m, y_m)\}$ of
m independent interpretations, the goal is to learn a mapping $h : \mathcal{X} \rightarrow \mathcal{Y}$,
where $\mathcal{X}$ denotes the set of all points x_i^k in any point cloud interpretation i, with
$i \in \{1, \ldots, m\}$ and $\mathcal{Y}$ is the set of target atoms y_i^k. The pair $e^k = (x_i^k, y_i^k)$ is a
training example, where $k \in \{1, \ldots, n\}$ and n is the number of training instances

in the point cloud interpretation i. One training example e^k is, thus, a smaller interpretation, part of the larger point cloud interpretation, and corresponds to one point in the object point cloud. Given a new point in a point cloud interpretation we can use h to predict its target category category/2.

3.4   Graphicalization

Next, each interpretation x is converted into a bipartite graph G which introduces a vertex for each ground atom. Vertices correspond to either entities or relationships, but identifiers are removed. Edges connect entities with relationships. Figure 3(b) shows part of the graph mapped on a point cloud. The graph is the result of grounding the E/R diagram for a particular point cloud.

4   Relational Kernel Features

We solve the grasping recognition problem in a supervised learning setting. We employ two variants of the fast neighborhood subgraph pairwise distance kernel [7]. The kernel is a decomposition kernel [13] that counts the number of common "parts" between two graphs. In our case the graph represents the contextual shape information of one point in the point cloud. The decomposition kernel between two graphs is defined with the help of relations $R_{r,d}$ ($r = 0, \ldots, R$ and $d = 0, \ldots, D$) as follows:

$$K(G, G') = \sum_{r=0}^{R} \sum_{d=0}^{D} \sum_{\substack{A, B \in R_{r,d}^{-1}(A, B, G) \\ A', B' \in R_{r,d}^{-1}(A', B', G')}} \kappa((A, B), (A', B')) \tag{1}$$

where $R_{r,d}^{-1}(A, B, G)$ returns the set of all pairs of neighborhoods (or balls) (A, B) of radius r with roots at distance d that exist in G. Thus, a "part" is a pair of neighborhoods (or a pair of balls). Figure 4 shows a neighborhood-pair feature with $R = 2$ and $D = 2$ for our grasping problem. The kernel hyper-parameters maximum radius R and maximum distance D are set experimentally. We ensure that only neighborhoods centered on the same type of vertex will be compared, constraint imposed by the equation:

$$\kappa((A, B), (A', B')) = \kappa_{root}((A, B), (A', B')) \cdot \kappa_{subgraph}((A, B), (A', B')), \tag{2}$$

where the component $\kappa_{root}((A, B), (A', B'))$ is 1 if the neighborhoods to be compared have the same type of roots, while the component $\kappa_{subgraph}((A, B), (A', B'))$ compares the pairs of neighborhood graphs extracted from two graphs G and G'. We solve the grasping problem using two specializations of $\kappa_{subgraph}$. Because we deal both with symbolic and numerical attributed graphs, we employ a hard-soft variant which combines an exact matching kernel for the symbolic relations and a soft match kernel for numerical properties of the relations, and a soft variant which uses only a soft match kernel.

Soft Matching. The soft matching kernel uses the idea of multinomial distribution (i.e., histogram) of labels. It discards the structural information inside the graph. Contextual information is still incorporated by the (sum) pooling operation applied on the numerical properties of the points.

$$\kappa_{subgraph}((A,B),(A',B')) = \sum_{\substack{v \in V(A) \cup V(B) \\ v' \in V(A') \cup V(B')}} \mathbf{1}_{\ell(v)=\ell(v')} \kappa_{tuple}(v,v') \qquad (3)$$

where $V(A)$ is the set of vertices of A and $\ell(v)$ is the label of vertex v. If the atom $point(p_1, f_1, \ldots, f_c, \ldots, f_m)$ is mapped into vertex v, $\ell(v)$ returns the signature name *point*. In this case κ is decomposed in a part that counts the vertices that share the same labels $\ell(v)$ in the neighborhood pair and ensures matches between tuples with the same signature name $(\mathbf{1}_{\ell(v)=\ell(v')})$, and a second part that takes into account the tuple of property values. These are real values and thus, the kernel on the tuple considers each element of the tuple independently with the standard product:

$$\kappa_{tuple}(v,v') = \sum_c prop_c(v) \cdot prop_c(v') \qquad (4)$$

where for the atom $point(p_1, f_1, \ldots, f_c, \ldots, f_m)$, mapped into vertex v, $prop_c(v)$ returns the property value f_c. In words, the kernel will count the number of symbolic labels and will sum property values that belong to vertices with same labels $l(v)$ that are contained in the neighborhood pair.

Hard-Soft Matching. The hard-soft variant replaces the label $l(v)$ in Eq. 3 with a relabeling procedure for the discrete signature names. We proceed with a canonical encoding that guarantees that each vertex receives a label that identifies it in the neighborhood graph based on the exact extracted structure of the ball with respect to the relabeled vertex. Then, the exact match kernel for the discrete part is defined as $\kappa_{subgraph}((A,B),(A',B')) = 1$ iff (A,B) and (A',B') are pairs of isomorphic graphs. The isomorphism is ensured by the vertices canonical relabeling. This match ensures that the contextual structure of the subgraphs matched is the same. Concerning the real valued properties, we use the standard product as in Eq. 4 for the tuples of vertices with the same relabelings. The spatial relations injected in the graph and its structure ensure that the pooled features are the ones belonging to vertices with a similar relabeling. In this way, we only sum the features with same contextual structure.

There are several advantages of using kLog and its kernel-based language. First, it can take relational contextual features into account in a principled way. Second, it allows fast computations with respect to the interpretation size, which allows us to explore different measures of contextual information via the kernel hyper-parameters. Third, it provides a flexible architecture in which only the specification language for relational learning problems is fixed. Actual features are determined by the choice of the graph kernel. In this setting, experimenting with alternative feature spaces is rapid and intuitive. For more details, see [11].

5   Experiments

We evaluate whether our relational kernel-based approach can exploit contextual shape information by pooling numerical features. Specifically, we investigate the following questions:

(Q1) Does numerical shape feature pooling improve upon local shape features for the robot grasping task considered?

(Q2) Does hard-soft matching improve over soft matching when incorporating contextual shape information?

To answer these questions, we perform experiments with all shape features considered in turn.

5.1   Dataset and Evaluation

We consider a realistic dataset similar to that in [23]. It is gathered using 8 objects: ellipsoidal object, rectangular object, round object, 2 glasses and 3 cups. The dataset contains 2631 instances (1972 positives and 659 negatives) and it was obtained in the ORCA simulation environment [3]. To gather the dataset, the robot performed grasping trials on a large number of reaching points. The setup is shown in Fig. 1 (left).

The goal is to evaluate the performance of our approach across the different objects considered. We estimate it under partial views, that is, each object is characterized by several partial point clouds, one for each view. The number of views can differ from object to object. Figure 5 shows four views for one of the cups. In practice, all views belonging to the same object are mapped to one interpretation, and thus, one interpretation corresponds to one object. Because the views are not spatially aligned, we consider spatial relations only between points that belong to the same view.

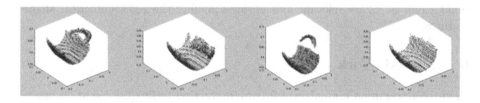

Fig. 5. Point clouds representing partial views of a cup.

For performance evaluation, we apply the leave-one-out CV method where one object is used for testing and the rest for training. In all our experiments we used a SVM with a linear kernel on top of the relational kernel features. The SVM cost parameter was set to 1. Because the dataset is unbalanced (with more positives than negatives), we evaluate performance in terms of the area under the ROC curve (AUC) and the area under the precision-recall curve (AP) which are not sensitive to the distribution of instances to classes. We also report the true positive rate (TPR) and accuracy (Acc) for both datasets. In order to better cope with the unbalanced data, the SVM implementation used (LIBSVM [6]) assigns different weights to positive and negative instances. In our case, we assign more weight to the negatives. The weight is selected using the leave-one-out CV for each feature type (when no relations are used), and is kept the same for that feature as we gradually add relations.

5.2   Results and Discussion

In the following we present quantitative experimental results for both questions[1]. Results in bold font indicate the best performance. For each feature type we start with local feature vectors and we gradually add the different relations considered, closeBy2/2 and closeBy3/3, respectively. As a baseline for comparison we use the local feature vectors alone, without any spatial relations. We also present results with all available features (VFH + PFH + SC = *all*) in one experiment with and without relations. We report performance results using the hard-soft matching kernel in Table 1 for sphere features and cell features setups. They are obtained for hyper-parameters $R=2$, $D=6$ (for *shape feature+closeBy2* settings) or $R=2$, $D=8$ (for the rest of the settings). The results in italics mark the best results for each local feature type (i.e., VFH/PFH/SC/all) in each grasping settings (gripper cell/sphere). The results in bold mark the best results for each grasping setting across all feature types considered. They show that the use of qualitative relations to pool features improves robot grasping performance for all shape feature types considered. This answers positively (Q1).

We answer question (Q2) by plotting the ROC curves for both soft and hard-soft kernels for sphere and cell features. The results in Fig. 6(a) and (b) show that hard-soft matching improves considerably upon soft matching. The curves correspond to hyper-parameters $R = 2$, $D = 8$ and closeBy2/2 + closeBy/3 relation, which give the best performance. Thus, contextual structure in the point cloud is highly relevant and ensures pooling the right numerical shape features.

6   Related Work

In visual recognition a number of feature extraction techniques based on image descriptors (e.g., SIFT) have been proposed. They usually encode the descriptors over a learned codebook and then summarize the distribution of the codes by a pooling step [5,14]. While the coding step produces representations that can be aggregated without losing too much information, pooling these codes gives robustness only to small transformations of the image. One fact that makes the coding step necessary in standard computer vision tasks is that image descriptors such as SIFT cannot be pooled directly with their neighbours without losing

---

[1] These results contain an errata to the results reported in [20].

Table 1. Hard-soft matching results for sphere and gripper cell setups.

Shape features	AUC		AP		Acc (%)		TPR (%)	
	sphere	cell	sphere	cell	sphere	cell	sphere	cell
VFH	0.66	0.70	0.83	0.86	60.43	75.33	58.52	*89.50*
VFH+closeBy2	0.80	0.83	0.92	0.93	72.33	80.16	73.33	89.45
VFH+closeBy3	0.83	0.84	*0.93*	*0.94*	78.91	81.45	82.05	88.74
VFH+closeBy2+closeBy3	*0.84*	*0.85*	*0.93*	*0.94*	*79.32*	*81.91*	*82.51*	89.35
PFH	0.70	0.71	0.88	0.86	60.62	76.24	56.80	91.13
PFH+closeBy2	0.80	0.83	0.92	0.93	73.20	79.82	74.29	90.11
PFH+closeBy3	0.82	0.85	*0.93*	*0.94*	77.77	82.25	80.68	91.08
PFH+closeBy2+closeBy3	*0.83*	**0.86**	*0.93*	*0.94*	*77.99*	**83.01**	*81.03*	*91.63*
SC	0.75	0.72	0.88	0.85	73.58	69.14	78.85	71.35
SC+closeBy2	0.80	0.80	0.92	0.90	76.09	77.27	81.29	82.86
SC+closeBy3	*0.83*	*0.81*	*0.93*	*0.91*	79.29	78.94	84.08	83.92
SC+closeBy2+closeBy3	*0.83*	*0.81*	*0.93*	*0.91*	**79.55**	*79.82*	**84.13**	*85.14*
all	0.75	0.71	0.89	0.86	74.15	76.66	80.02	92.55
all+closeBy2	0.81	0.84	*0.93*	0.94	74.15	81.53	78.35	**92.60**
all+closeBy3	*0.83*	**0.86**	*0.93*	**0.95**	78.72	82.25	83.06	91.99
all+closeBy2+closeBy3	*0.83*	**0.86**	*0.93*	**0.95**	*79.29*	*82.82*	*83.92*	92.29

information. Differently, our contribution for robot grasping considers shape feature pooling without the coding step, by means of SRL techniques.

Previous works on visual-dependent robot grasping have shown promising results on learning grasping points from image-based 2D descriptors [21,29]. Other works exploit combinations of image-based and point cloud-based features [4,15]. Saxena et. al. [29] propose to infer grasping probabilities from image filter responses at the object points. Their approach allows to discriminate graspable from non-graspable points and transfer knowledge to new objects. However, it does not consider the parameters of the gripper to estimate the quality of the grasping. Jiang et. al. [15] extend this approach by computing grasping stability features from the point clouds. In their method, the point cloud features are linked to the gripper configuration, while the image-based features are linked to the visual graspability of a point. Differently, we consider dense 3D data for both gripper configuration and visual graspability. Kraft et. al. [16,17] propose to learn by exploration graspable points of an object. Nevertheless, their learning procedure is specific to each object, and it is difficult to transfer the skills learned to other objects. A major difference is that we learn with features that generalize across objects.

Furthermore, a significant number of vision-based grasping methods learn mappings from 2D/3D features to grasping parameters [4,19,22,30]. However, it turns out that it is difficult to link a 3D gripper orientation to local shape features

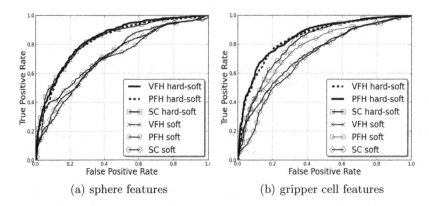

(a) sphere features   (b) gripper cell features

Fig. 6. ROC curves for soft and hard-soft matching kernels; $R=2, D=8$; VFH/PFH/ SC + closeBy2 + closeBy3.

without considering contextual or global object information. Only recently, methods that take global and symbolic information into account have been proposed [1]. They benefit from increased geometric robustness, which gives advantages with respect to the pre-shape of the robotic hand and general shape of the object, generating more accurate grasps. Nevertheless, this work relies on complete object point clouds, and object reconstruction based on single views is a difficult problem due to lack of observability of the self-occluded parts. Differently, our contribution to robot grasping exploits contextual shape information of objects from partial views and, additionally, we employ a new relational approach to vision-based grasping that considers symbolic and numerical attributed graphs.

From the SRL perspective, purely relational learning techniques have been previously used to learn from point clouds. The work in [9,10] uses first-order clause inducing systems to learn from discrete primitives (e.g., planes, cylinders) classifiers for concepts such as boxes, walls, cups or stairs. Differently, we propose a SRL approach to recognize graspable points that is based on relational kernels. A related graph kernel designed for classification and retrieval of labeled graphs was employed in [25,26]. There, in the context of robot grasping, the authors consider the tasks of object categorization and similar object view retrieval. Object graphs are obtained as k-nearest neighbor graphs from object point clouds and graph nodes are characterized by semantic labels. The kernel is an evolving propagation kernel based on continuous distributions as graph features, which are built from semantic node labels, and on a locality sensitive hashing function to ensure meaningful features. In contrast, our work focuses on recognizing graspable points in the cloud by employing a flexible decomposition kernel. It takes as input numerical shape vectors organized in graph structures, and computes graph features by pooling meaningful shape features that are ensured by the structure of the graph. In this case, we construct object graphs declaratively using relational background knowledge and we characterize graph nodes by numerical shape features instead of semantic labels. A similar graph kernel was employed in [2] for visual recognition of houses. However, there, the visual input features were discrete and did not have a continuous numerical form.

7   Conclusions

This paper proposes a relational kernel-based approach to recognize graspable object points. We represent each object as an attributed graph, where nodes corresponding to object points are characterized by distributions of numerical shape features. Extended contextual object shape information is encoded via qualitative spatial relations among object points. Next, we use kernels on graphs to compute highly discriminative features based on contextual information. We show experimentally that pooling spatially related numerical shape feature improves robot grasping results upon purely local shape-based approaches.

We point out three directions for future work. A first direction is to investigate how similar SRL techniques working directly with numerical features can help other robot vision tasks. A second direction is to validate our results on datasets that contain a wider range of object categories. Finally, a third direction is to investigate other spatial relations or domain knowledge that could give even better results for the robot grasping problem considered.

References

1. Aleotti, J., Caselli, S.: Part-based robot grasp planning from human demonstration. In: ICRA, pp. 4554–4560 (2011)
2. Antanas, L., Frasconi, P., Costa, F., Tuytelaars, T., De Raedt, L.: A relational kernel-based framework for hierarchical image understanding. In: Gimel'farb, G.L., Hancock, E.R., Imiya, A.I., Kuijper, A., Kudo, M., Shinichiro Omachi, S., Windeatt, T., Yamada, K. (eds.) SSPR&SPR 2012. LNCS, vol. 7626, pp. 171–180. Springer, Heidelberg (2012)
3. Baltzakis, H.: Orca simulator (2005). http://www.ics.forth.gr/cvrl/_software/orca_setup.exe
4. Bohg, J., Kragic, D.: Learning grasping points with shape context. RAS **58**(4), 362–377 (2010)
5. Boureau, Y.L., Bach, F., LeCun, Y., Ponce, J.: Learning mid-level features for recognition. In: CVPR, pp. 2559–2566 (2010)
6. Chang, C., Lin, C.: LIBSVM: a library for support vector machines. ACM Trans. Intell. Syst. Technol. **2**(3), 27:1–27:27 (2011)
7. Costa, F., De Grave, K.: Fast neighborhood subgraph pairwise distance kernel. In: ICML, pp. 255–262 (2010)
8. De Raedt, L.: Logical and Relational Learning. Cognitive Technologies. Springer, New York (2008)
9. Farid, R., Sammut, C.: Plane-based object categorisation using relational learning. ML **94**(1), 3–23 (2014)

10. Farid, R., Sammut, C.: Region-based object categorisation using relational learning. In: Pham, D.-N., Park, S.-B. (eds.) PRICAI 2014. LNCS, vol. 8862, pp. 357–369. Springer, heidelberg (2014)
11. Frasconi, P., Costa, F., De Raedt, L., De Grave, K.: kLog: a language for logical and relational learning with kernels. Artif. Intell. **217**, 117–143 (2014)
12. Garcia-Molina, H., Ullman, J.D., Widom, J.: Database Systems: The Complete Book, 2nd edn. Prentice Hall Press, Upper Saddle River (2008)
13. Haussler, D.: Convolution kernels on discrete structures. Technical report UCSC-CRL-99-10, University of California at Santa Cruz (1999)
14. Jia, Y., Huang, C., Darrell, T.: Beyond spatial pyramids: Receptive field learning for pooled image features. In: CVPR, pp. 3370–3377 (2012)
15. Jiang, Y., Moseson, S., Saxena, A.: Efficient grasping from rgbd images: Learning using a new rectangle representation. In: ICRA, pp. 3304–3311 (2011)
16. Kraft, D., Detry, R., Pugeault, N., Baseski, E., Guerin, F., Piater, J.H., Krüger, N.: Development of object and grasping knowledge by robot exploration. IEEE T. Auton. Mental Dev. **2**(4), 368–383 (2010)
17. Kraft, D., Detry, R., Pugeault, N., Başeski, E., Piater, J., Krüger, N.: Learning objects and grasp affordances through autonomous exploration. In: Fritz, M., Schiele, B., Piater, J.H. (eds.) ICVS 2009. LNCS, vol. 5815, pp. 235–244. Springer, Heidelberg (2009)
18. Krtgen, M., Novotni, M., Klein, R.: 3D shape matching with 3D shape contexts. In: The 7th Central European Seminar on Computer Graphics (2003)
19. Lenz, I., Lee, H., Saxena, A.: Deep learning for detecting robotic grasps. CoRR abs/1301.3592 (2013)
20. Mocanu-Antanas, L.: Relational Visual Recognition. Ph.D. thesis, Informatics Section, Department of Computer Science, Faculty of Engineering Science (2014)
21. Montesano, L., Lopes, M.: Learning grasping affordances from local visual descriptors. In: ICDL, pp. 1–6. IEEE Computer Society (2009)
22. Montesano, L., Lopes, M.: Active learning of visual descriptors for grasping using non-parametric smoothed beta distributions. Humanoids **60**(3), 452–462 (2012)
23. Moreno, P., Hornstein, J., Santos-Victor, J.: Learning to grasp from point clouds. Technical report Vislab-TR001/2011, Department of Electrical and Computers Engineering, Instituto Superior Técnico, Portugal, September 2011
24. Muja, M., Ciocarlie, M.: Table top segmentation package (2012). http://www.ros.org/wiki/tabletop_object_detector
25. Neumann, M., Garnett, R., Moreno, P., Patricia, N., Kersting, K.: Propagation kernels for partially labeled graphs. In: MLG-2012 (2012)
26. Neumann, M., Moreno, P., Antanas, L., Garnett, R., Kersting, K.: Graph kernels for object category prediction in task-dependent robot grasping. In: MLG-2013 (2013)
27. Rusu, R.B.: Semantic 3D Object Maps for Everyday Manipulation in Human Living Environments. Ph.D. thesis, Computer Science Department, Technische Universitat Munchen, Germany, October 2009
28. Rusu, R.B., Bradski, G., Thibaux, R., Hsu, J.: Fast 3D recognition and pose using the viewpoint feature histogram. In: IROS. Taipei, Taiwan, October 2010
29. Saxena, A., Driemeyer, J., Ng, A.Y.: Robotic grasping of novel objects using vision. IJRR **27**(2), 157–173 (2008)
30. Saxena, A., Wong, L.L.S., Ng, A.Y.: Learning grasp strategies with partial shape information. In: AAAI, pp. 1491–1494. AAAI Press (2008)

CARAF: Complex Aggregates within Random Forests

Clément Charnay$^{(\boxtimes)}$, Nicolas Lachiche, and Agnès Braud

ICube, Université de Strasbourg, CNRS, 300 Bd Sébastien Brant - CS 10413,
67412 Illkirch Cedex, France
{charnay,nicolas.lachiche,agnes.braud}@unistra.fr

Abstract. This paper presents an approach integrating complex aggregate features into a relational random forest learner to address relational data mining tasks. CARAF, for Complex Aggregates within RAndom Forests, has two goals. Firstly, it aims at avoiding exhaustive exploration of the large feature space induced by the use of complex aggregates. Its second purpose is to reduce the overfitting introduced by the expressivity of complex aggregates in the context of a single decision tree. CARAF compares well on real-world datasets to both random forests based on the propositionalization method RELAGGS, and the relational random forest learner FORF. CARAF allows to perform complex aggregate feature selection.

1 Introduction and Context

Relational data mining, as opposed to attribute-value learning, refers to learning from data represented across several tables. These tables represent different objects, linked by relationships. Many datasets from many domains fall into the relational paradigm, leading to a much richer representation. The applications go from the molecular domain, to geographical data, and any kind of spatio-temporal data such as speech recognition.

The difference to attribute-value learning is the one-to-many relationship. In particular, we focus on a two-table setting: one table, the main table, represents the objects we want to perform prediction on. This prediction, supervised learning, task is either a classification task if the attribute to predict is categorical, i.e. if it takes a finite number of values, or a regression task, is the attribute to predict is numeric. The second table, referred to as the secondary table, contains objects related to the main ones in a one-to-many relationship, which means several secondary objects are linked to one main object. In practice, many datasets are represented in this two-table setting: sequential data is represented as a main table containing information on the sequence, while the secondary table contains the elements of the sequence. The multi-dimensional setting is another use case, where one is often interested in learning on one dimension based on the contents of the table of facts, which are linked through a one-to-many relationship.

As an example, the relational schema for the Auslan dataset, an Australian sign language recognition task, is given in Fig. 1. It is a classification task, where the aim

© Springer International Publishing Switzerland 2016
K. Inoue et al. (Eds.): ILP 2015, LNAI 9575, pp. 15–29, 2016.
DOI: 10.1007/978-3-319-40566-7_2

is to predict the sign associated to a record of hand motion. The main table, associated to records, contains only the attribute to learn, i.e. the language sign associated to the record, while the secondary table contains the samples of the records, with a timestamp attribute and 22 attributes representing values from the channels that monitor the hand motion through a glove.

1

0..N

record_id	sign
1	alive
2	all
3	answer
...	...

Main table - records

element_id	record_id	time	chan1	...	chan22
1_1	1	1	-0.1	...	0.09
1_2	1	2	-0.08	...	0
1_3	1	3	-0.06	...	1
...	...	...	...	...	...
2_1	2	1	0.02	...	1
2_2	2	2	0.02	...	0.9
...	...	...	...	...	...

Secondary table - elements

Fig. 1. Schema of the real-world Auslan dataset.

Most relational data mining algorithms are based on inductive logic programming concepts, and handle the relationships through the use of the existential quantifier: it introduces a secondary object B linked to the main object A, B usually meets a certain condition and its existence is relevant to classify A. For instance, on the Auslan dataset, to discriminate between signs, a feature like the fact that an element of the record has a value higher than 0.9 for channel 13 could be useful. TILDE [2] is a relational extension of Quinlan's C4.5 [11] decision tree learner based on this idea. Other approaches use aggregates: they take all B objects linked to A, and aggregate the set to one value, for instance by computing the average of a numeric property of the B objects. For instance, the average value of channel 9 over the whole record may help discriminate between signs. The propositionalization approach RELAGGS [9] introduces such aggregates.

One approach combines both, by filtering the B objects on a condition before aggregating them. This approach is known as complex aggregation. As opposed to simple aggregation, it consists in aggregating a subset of the B objects linked to A, the subset being defined by a conjunction of conditions over the attributes of the secondary table. For instance, a feature that may be useful to classify signs could be the average value of channel 9 over record elements between timestamps 15 and 22.

However, the complex aggregates introduce two specific challenges: firstly, the introduction of a condition prior to the aggregation increases exponentially the size of the search space, which makes an exhaustive exploration intractable. Secondly, the complex aggregates, being a very rich representation, are also very specific and strict, which implies they are prone to overfitting. Especially, complex aggregate-based algorithms consider also the simple aggregates that RELAGGS builds. Therefore, when a model introduces complex aggregates, it means they have been found to increase the performance on the training set over simple aggregates. If on test data, the complex aggregate-based model performs worse than a simple aggregate-based one, there is overfitting.

In this paper, we propose the extension of the decision tree learner based on RRHCCA to a random forest learner, introducing two faster hill-climbing algorithms. The method to perform complex aggregate feature selection in order to identify relevant families of aggregates is also presented.

The rest of the paper is organized as follows: in Sect. 2 we briefly define the concept of complex aggregates. In Sect. 3, we review the use of random forests in the relational setting. In Sect. 4, we introduce CARAF (Complex Aggregates with RAndom Forests), a new relational random forest learner implementing our contributions. In Sect. 5, we present experimental results obtained with CARAF. In Sect. 6, we present how to perform complex aggregate feature selection with CARAF. Finally, in Sect. 7, we conclude and give some future work perspectives.

2 Complex Aggregates

In this section, we briefly define the concept of complex aggregates, which has been thoroughly explained in [5].

In a setting with two tables linked through a one-to-many relationship, let us denote the main table by M and the secondary table by S. We define a complex aggregate feature of table M as a triple *(Selection, Feature, Function)* where:

- Selection selects the objects to aggregate. It is a conjunction of s conditions, i.e. $Selection = \bigwedge_{1 \leq i \leq s} c_i$, where c_i is a condition on a descriptive attribute of the secondary table. Formally, let $S.A$ be the set of descriptive attributes of table S, and $Attr \in S.A$, then c_i is:
  - either $Attr \in vals$ with *vals* a subset of the possible values of *Attr* if *Attr* is a categorical feature,
  - or $Attr \in [val_1; val_2[$ if *Attr* is a numeric feature.
  In other words, for a given object of the main table, the objects of the secondary table that meet the conditions in *Selection* are selected for aggregation.
- *Feature* can be:
  - nothing,
  - a descriptive attribute of the secondary table, i.e. $Feature \in S.A$.
  Thus, *Feature* is the attribute of the selected objects that will be aggregated. It can be nothing since the selected objects can simply be counted, in which case a feature to aggregate is not needed.
- *Function* is the aggregation function to apply to the bag of feature values for the selected objects. Aggregation functions we consider are *count* which does not need an attribute to aggregate, *min, max, sum, average, standard deviation, median, first quartile, third quartile, interquartile range, first decile and ninth decile* for numeric attributes, *proportion of secondary objects with*

attribute value for categorical attributes, the latter is defined as the ratio of secondary objects linked to a given main object with a given value for the attribute, over the count of all secondary objects linked this main object.

In the rest of the paper, we will denote a complex aggregate by *Function(Feature, Selection)*. We will refer to the set of possible *(Function, Feature)* pairs as the aggregation processes, i.e. the different possibilities to aggregate a set of secondary objects.

The introduction of a condition on the objects to aggregate makes the feature space impossible to explore exhaustively. Heuristics have been proposed to explore this space in a smart way. The refinement cube [14] is based on the idea of the monotonicity of the dimensions of the cube. Indeed, the aggregation condition, aggregation function and threshold can be explored in a general-to-specific way, using monotone paths: when a complex aggregate (a point in the refinement cube) is too specific (i.e. it fails for every training example), the search does not restart from this point. The approach introduced in [3] builds complex aggregate features for use in a Bayesian classifier, guided by minimum description length and a heuristic sampling. The RRHCCA algorithm [5] has been proposed to explore a larger search space with a random-restart hill-climbing approach to find the appropriate condition with respect to the aggregation process, still in the context of a decision tree learner. However, the decision tree model with complex aggregates often fails to outperform RELAGGS, which shows overfitting. As a solution, we propose its extension to a Random Forest model.

3   Random Forests

Random Forest [4] is an ensemble classification technique which builds a set of diverse decision trees and combines their predictions into a single output. Considered individually, each decision tree is less accurate than a decision tree built in a classic way. However, the introduction of diversity through the forest improves the performance over a single decision tree, by solving the overfitting problem induced by the latter approach. Algorithm 1 shows the building process of a Random Forest. Diversity between the trees is achieved by two means:

- Bootstrapping: each tree is built on a different training set using sampling with replacement from the original training set. In other words, each decision tree is built using a training set with same size as the original one, but where repetitions may occur. This corresponds to lines 5 to 8 of Algorithm 1.
- Feature sampling: to build each node of each tree, a subset of features is used. If there are *numFeatures* available, $\sqrt{numFeatures}$ are considered for introduction in node split. This corresponds to lines 9 to 15 of Algorithm 1.

The use of Random Forests for relational data mining purposes is not new: TILDE decision trees have been used as a basis for FORF (First-Order Relational Random Forests) [13], which can, as TILDE, be used with complex aggregates. However, the implementation suffers memory limitations, e.g. allocation failures

---

Algorithm 1. BuildRandomForest

1: **Input:** *train*: set of training examples, *feats*: set of possible split features, *target*:
the target attribute, *n*: number of trees in the forest.
2: **Output:** *forest*: a random forest.

3: forest ← InitEmptyForest()
4: **for** k = 1 **to** n **do**
5:     trainForTree ← InitEmptyInstances()
6:     **for** i = 1 **to** train.Size() **do**
7:         trainForTree.Add(train.OneRandomElement())
8:     **end for**
9:     featsCopy ← feats.Copy()
10:     featsForTree ← InitEmptyFeatures()
11:     **for** j = 1 **to** $\sqrt{feats.Size()}$ **do**
12:         f ← featsCopy.OneRandomElement()
13:         featsCopy.Remove(f)
14:         featsForTree.Add(f)
15:     **end for**
16:     tree ← BuildDecisionTree(trainForTree, featsForTree, target)
17:     forest.Add(tree)
18: **end for**
19: **return** forest

---

when the feature space induced by the language bias is too wide. Also, the logic programming formalism makes the case of empty sets ambiguous. Indeed, the failure of a comparison test on an aggregate can have two reasons: the comparison can actually fail or the aggregate predicate can fail because it cannot compute a result, generally because the set to aggregate is empty. In the implementation of CARAF, we overcome this limitation by considering aggregation failure as a third outcome of a test.

Another relational Random Forest algorithm is described in [1]. It uses random rules based on the existential quantifier. However, it does not consider aggregates.

4   CARAF: Complex Aggregates with RAndom Forests

In this section, we describe the main contributions brought by CARAF (Complex Aggregates with RAndom Forests).

First is the use of random forests. The instance bootstrapping part is performed the same way as Breiman does, by sampling with replacement from the training set. The feature sampling is different, based on the complex aggregates space structure. Let us denote by $AggProc = |(Function, Feature)|$ the number of aggregation processes, N_s the number of secondary objects, and A the number of attributes in the secondary table. The number of conjunctions of conditions, i.e. the number of possible *Selection* grows like N_s^A for numeric attributes. A good estimation for the number of complex aggregates is then

$ComplAgg = AggProc \cdot N_s^A$. As a subsampling method, we want to keep a search space of size $\sqrt{ComplAgg}$. We then keep $\sqrt{AggProc}$ aggregation processes and, in each process, $A/2$ attributes to put conditions on. This gives us the desired feature subsampling.

For instance, let us consider again the Auslan dataset. For sake of simplicity, we consider *count, minimum, maximum and average* as the possible aggregation function, and attributes time and channels 1 to 4. Table 1 shows an example of complex aggregates subsampling on this dataset. Out of the 16 aggregation processes available, the square root will be considered at each node, i.e. 4, as shown in Table 1a. For each aggregation process, half of the 5 secondary attributes will be kept for use in the selection conjunction of conditions, i.e. 3 per aggregation process, as shown in Table 1b.

Table 1. Subsampling of complex aggregates.

(a) Subsampling of aggregation processes.

Function	Attribute	Chosen
Count		x
Minimum	Time	
Minimum	Chan1	
Minimum	Chan2	
Minimum	Chan3	
Minimum	Chan4	
Maximum	Time	
Maximum	Chan1	
Maximum	Chan2	x
Maximum	Chan3	
Maximum	Chan4	
Average	Time	x
Average	Chan1	
Average	Chan2	
Average	Chan3	
Average	Chan4	x

(b) Subsampling of secondary attributes.

Attribute	Chosen
Time	x
Chan1	x
Chan2	
Chan3	x
Chan4	

The RRHCCA algorithm aims at exploring the complex aggregates search space in a stochastic way. It uses random restart hill-climbing to find the best conjunction of conditions *Selection* for a given aggregation process *(Function, Feature)*. The hill-climbing process used to search this space can be RRHCCA, but we chose to simplify it to make it less time-consuming. We propose two approaches to achieve that.

We first introduce the "Random" hill-climbing algorithm, for which pseudo-code is given in Algorithm 2. Like RRHCCA, the aim is to look for an appropriate conjunction of basic conditions for a fixed aggregation process. But instead of considering all neighbors of an aggregate at each step of hill-climbing, the Random algorithm will consider only one, randomly chosen, neighbor, for split evaluation.

Algorithm 2. Random Hill-Climbing Algorithm

1: **Input:** *functions*: list of aggregation functions, *features*: list of attributes of the secondary table, *train*: labelled training set.
2: **Output:** *split*: best complex aggregate found through hill-climbing.

3: aggregationProcesses ← InitializeProcesses(functions, features)
4: bestSplits ← []
5: bestScore ← WORST_SCORE_FOR_METRIC
6: **for all** aggProc ∈ aggregationProcesses **do**
7:    iterWithoutImprovement ← 0
8:    **for** i = 1 **to** MAX_ITERATIONS **and** iterWithoutImprovement < 0.2*MAX_ITERATIONS **do**
9:       hasImproved ← aggProc.GrowRandom(train)
10:      **if not** hasImproved **then**
11:         iterWithoutImprovement++
12:         **if** aggProc.split.score ≥ bestScore **then**
13:            **if** aggProc.split.score > bestScore **then**
14:               bestScore ← aggProc.split.score
15:               bestSplits ← []
16:            **end if**
17:            bestSplits.Add(aggProc.split)
18:         **end if**
19:      **else**
20:         iterWithoutImprovement ← 0
21:      **end if**
22:   **end for**
23: **end for**
24: split ← bestSplits.OneRandomElement()
25: **return** split

Algorithm 3. AggregationProcess.GrowRandom: Perform One Step of Hill-Climbing for the Aggregation Process

---

1: **Input:** *train*: labelled training set.
2: **Output:** *hasImproved*: boolean indicating if the step of the hill-climbing has improved the best split found in the current hill-climbing of the aggregation process.

3: allNeighbors ← EnumerateNeighbors(this.aggregate.condition)
4: neighbor ← allNeighbors.OneRandomElement()
5: aggregateToTry ← CreateAggregate(this.aggregate.function, this.aggregate.feature, neighbor)
6: spl ← EvaluateAggregate(aggregateToTry, train)
7: hasImproved ← UpdateBestSplit(spl)
8: **return** hasImproved

Algorithm 4. Global Hill-Climbing Algorithm

1: **Input:** *functions*: list of aggregation functions, *features*: list of attributes of the secondary table, *train*: labelled training set.
2: **Output:** *split*: best complex aggregate found through hill-climbing.

3: aggregationProcesses ← InitializeProcesses(functions, features)
4: bestSplits ← []
5: bestScore ← WORST_SCORE_FOR_METRIC
6: conjunction ← InitEmptyConjunction()
7: iterWithoutImprovement ← 0
8: **for** i = 1 **to** MAX_ITERATIONS **and** iterWithoutImprovement < 0.2*MAX_ITERATIONS **do**
9:     allNeighbors ← EnumerateNeighbors(conjunction)
10:    neighbor ← allNeighbors.oneRandomElement()
11:    hasImproved ← false
12:    **for all** aggProc ∈ aggregationProcesses **do**
13:        aggregateToTry ← CreateAggregate(aggProc.function, aggProc.feature, neighbor)
14:        spl ← EvaluateAggregate(aggregateToTry, train)
15:        **if** spl.score ≥ bestScore **then**
16:            **if** spl.score > bestScore **then**
17:                bestScore ← spl.score
18:                bestSplits ← []
19:                hasImproved ← true
20:            **end if**
21:            bestSplits.Add(spl)
22:        **end if**
23:    **end for**
24:    **if** hasImproved **then**
25:        iterWithoutImprovement ← 0
26:    **else**
27:        iterWithoutImprovement++
28:    **end if**
29: **end for**
30: split ← bestSplits.OneRandomElement()
31: **return** split

This corresponds to the function GrowRandom shown in Algorithm 3. If the chosen neighbor improves over the original aggregate, the search resumes from this neighbor. The neighbors are defined as in Algorithm 5: from a current aggregate, they are obtained by adding a random basic condition to the conjunction, removing a condition from the conjunction, and modifying one.

The hill-climbing has two possible stopping criteria: when a maximum number of hill-climbing steps have been performed, or when a certain number of neighbors of a given aggregate have been considered without improvement, this number has been arbitrarily fixed to 20 % of the maximum number of hill-climbing steps. In other words, if 20 % of the maximum number of iterations

---

Algorithm 5. EnumerateNeighbors

1: **Input:** *conjunction*: aggregation conjunction of conditions.
2: **Output:** *allNeighbors*: array of aggregation conjunctions, neighbors of *conjunction*.

3: allNeighbors ← []
4: **for all** attr ∈ secondary attributes not present in *conjunction* **do**
5:    nextConjunction ← conjunction obtained by adding one randomly initialized condition on attr to *conjunction*
6:    allNeighbors.Add(nextConjunction)
7: **end for**
8: **for all** attr ∈ secondary attributes already present in *conjunction* **do**
9:    nextConjunction ← condition obtained by removing the condition on attr present in *conjunction*
10:    allNeighbors.Add(nextConjunction)
11: **end for**
12: **for all** attr ∈ secondary attributes already present in *conjunction* **do**
13:    **for all** move ∈ possible moves on the condition on attr present in *conjunction* **do**
14:       nextConjunction ← aggregate obtained by applying move to *conjunction*
15:       allNeighbors.Add(nextConjunction)
16:    **end for**
17: **end for**
18: **return** allNeighbors

---

have passed with no improvement, the search stops. This aggregation process-wise hill-climbing loop corresponds to lines 8 through 22.

This hill-climbing search is then performed once for each aggregation process available, starting from an empty conjunction of conditions, without a restart. This corresponds to the loop from line 6 to line 23.

Following the idea of the Random hill-climbing algorithm, we propose to invert the loops of hill-climbing and aggregation process, materialized in the "Global" hill-climbing algorithm. In practice, only one hill-climbing search is performed, which aims at finding the best conjunction of conditions for all aggregation processes available. For a given conjunction of conditions, all aggregation processes are used to form aggregates and splitting conditions, and the conjunction is evaluated according to the best score achieved over all aggregation processes. The pseudo-code is given in Algorithm 4. This time, the aggregation process loop (from line 12 to line 23) is enclosed in the hill-climbing loop (from line 8 to line 29).

An additional feature is the use of ternary decision trees instead of binary decision trees. Each internal node of the tree has three sub-branches: one for success of the test, one for actual failure, and one for the unapplicability of the test, e.g. if the value of the feature involved in the test cannot be computed for the instance at hand. This is a way of dealing with empty sets in the context of complex aggregates. Indeed, imposing conditions on the secondary objects to aggregate can result in the absence of objects to be aggregated, i.e. aggregating an empty set. This is a problem for most aggregation functions, e.g. the average. We choose to tackle this issue by considering this as a third possible outcome of the test.

5   Experimental Results

In this section, we compare CARAF using the 3 different hill-climbing approaches to RELAGGS used in combination with Random Forest in Weka [8], and to FORF. All random forests were run to build 33 trees. We consider seven real-world real datasets.

- Auslan is a task of recognition of the Australian language sign.
- Diterpenes [7] is a molecule classification task.
- Japanese vowels is related to recognition of Japanese vowels utterances from cepstrum analysis.
- Musk1 and Musk2 [6] are molecule classification tasks.
- Mutagenesis [12] is about predicting mutagenicity of a molecule with respect to the properties of its atoms. In out two-table setting, we use the so-called "regression-friendly" subset of the dataset, and consider a molecule as a bag of atoms, i.e. we do not consider the bond information between atoms.
- Opt-digits deals with optical recognition of handwritten digits.
- Urban blocks [10] is a geographical classification task. This dataset is a clean version of the one used in [5] in the sense that duplicate urban blocks were removed.

A description of the datasets is given in Table 2.

The accuracy results are reported in Table 3. It is test set accuracy when a test set is available for the dataset or out-of-bag accuracy on the training set when there is no test set. Out-of-bag error is defined as follows: as mentioned previously, each tree in a Random Forest is trained using a subsample of the original training set, i.e. for each tree, there is a fraction of the training set that has not been actually used to build the tree. The out-of-bag accuracy for the tree is the error made by the tree on this set of unseen examples, called the out-of-bag examples. Any error metric can be used. For classification tasks, error rate will be most likely used, while for regression tasks root mean squared error could be used. By extension, out-of-bag accuracy is defined as the complementary to 1 of the out-of-bag classification error rate. The figures in bold indicate that the difference with RELAGGS is statistically significant with 95 % confidence, while the underlined figures indicate a significant difference with FORF. The run of FORF on the Auslan dataset resulted in an unknown error and cannot be reported.

We observe that CARAF with the original RRHCCA hill-climbing algorithm is always performing better than both RELAGGS and FORF, the difference being significant in 3 cases out of 8 over RELAGGS, and 4 out of 7 over FORF. The Random and Global hill-climbing approaches also perform better

Table 2. Characteristics of the datasets used in the experimental comparison.

Dataset	Instances	Classes	Secondary objects	Secondary attributes
Auslan	2 565	96	146 949	23
Diterpenes	1 503	23	30 060	2
Japanese vowels	270 + 370	9	9 961	12
Musk1	92	2	476	166
Musk2	102	2	6 598	166
Mutagenesis	188	2	4 893	3
Opt-digits	3 823 + 1 797	10	5 754 880	3
Urban blocks	591	6	7 692	3

Table 3. Results of CARAF with different hill-climbing heuristics on different datasets (out-of-bag accuracy or test set accuracy).

Dataset	RELAGGS	FORF	RRHCCA	Random	Global
Auslan	94.19 %	ERR	**96.53 %**	**95.91 %**	94.66 %
Diterpenes	89.09 %	90.49 %	**92.95 %**	85.06 %	**93.35 %**
Japanese vowels	93.78 %	94.86 %	95.41 %	**97.30 %**	**97.03 %**
Musk1	80.43 %	78.26 %	89.13 %	84.78 %	80.43 %
Musk2	76.47 %	75.49 %	81.37 %	85.29 %	82.35 %
Mutagenesis	88.30 %	87.77 %	90.43 %	91.49 %	92.02 %
Opt-digits	22.37 %	76.57 %	**95.94 %**	**94.60 %**	**92.77 %**
Urban blocks	83.42 %	75.81 %	84.94 %	83.76 %	84.60 %
			8 **(3)** - 7 (4)	7 **(3)** - 6 (2)	7.5 **(3)** - 7 (3)

than RELAGGS and FORF in a majority of cases, some cases also being statistically significant. These two approaches, considering less complex aggregates, also have the advantage of speed over RRHCCA. As shown in Table 4, the runtimes of both Random and Global are lower by a factor at least 4 than the runtimes of RRHCCA, Global being faster than Random. The loss in accuracy performance is tiny: RRHCCA outperforms Random 5 times, the difference being statistically significant only once. RRHCCA outperforms Global 4 times, significantly twice. The Random and Global approaches are then good performers too. Therefore, our recommendation is, if runtime is not a problem for the dataset at hand, to use RRHCCA. If time is critical, then Random is the best option, followed by Global.

Table 4. Runtime of the algorithms (in minutes).

Dataset	RRHCCA	Random	Global
Auslan	921	250	146
Diterpenes	4	1	1
Japanese vowels	13	1	1
Musk1	98	8	5
Musk2	733	71	55
Mutagenesis	6	2	2
Opt-digits	35	9	5
Urban blocks	4	1	1

6   Aggregation Processes Selection with Random Forests

Random Forests can be used to perform feature selection, as introduced by Breiman in [4]. The aim is to first check which families of complex aggregates are the most promising, to learn a model afterwards using only these useful families.

Our goal is to perform feature selection, i.e. to assess the importance of an input feature for prediction of the output attribute. This achieved using permutation tests. For a given tree, we first measure the out-of-bag error. The second step is to permute among the out-of-bag examples the value for the input feature we want to measure the importance. This gives a new out-of-bag examples set, for which we compute an after-permutation out-of-bag error. The importance of the feature at the tree-level is the increase in error between the after-permutation out-of-bag set and the original out-of-bag set. The final feature importance is then obtained by averaging tree-level feature importances over the whole forest.

In a relational context where complex aggregates are being used, this method needs adaptation. Indeed, the size of the complex aggregates search space implies that a given complex aggregate is rarely used twice in the same model. However, the structure of the complex aggregates allows us to define families of complex aggregates, and to measure importance of the families rather than specific complex aggregates.

Families of complex aggregates can be defined according to two elements:

- Aggregation processes: Complex aggregates sharing a common aggregation process will belong to the same family.
- Attributes in selection conjunctions: Complex aggregates whose selection conjunctions of conditions have a condition on a common attribute will belong to the same family.

These two elements can be combined to define more specific attributes, e.g. complex aggregates with the same aggregation process whose conjunctions of conditions have a condition on the same given attribute.

As an example, we use the urban blocks dataset from Sect. 5. We consider *count*, *minimum*, *maximum* and *average* as the possible aggregation functions. Block-wise features are area, elongation, convexity and density, while building-wise features are area, elongation and convexity.

If we define families of complex aggregates at the aggregation process level, we obtain as many families as aggregation processes, 10 in this example. Thus, following aggregates will fall into the same family, since they are all based on the same aggregation process, the average area of buildings:

- $average(area, buildings, true)$
- $average(area, buildings, elongation \geqslant 0.7)$
- $average(area, buildings, convexity < 0.5)$

If we define families based on one common attribute in the conjunction of conditions, we have as many families as attributes in the secondary table, 3 in this example. Thus, following aggregates will fall into the same family, since their conjunctions of conditions all have a condition on elongation of buildings:

- $average(area, buildings, elongation \geqslant 0.7)$
- $maximum(convexity, buildings, elongation < 0.6)$
- $count(buildings, elongation < 0.8 \land area \geqslant 100)$

Both can be combined to create families based on the aggregation process and a common attribute in conjunction of conditions, 30 in this example. For instance, following aggregates will belong to the same family, sharing both the aggregation process of average area of buildings and a condition on elongation of buildings:

- $average(area, buildings, elongation \geqslant 0.7)$
- $average(area, buildings, elongation < 0.9 \land convexity \geqslant 0.7)$
- $average(area, buildings, elongation \geqslant 0.5 \land area < 100)$

The permutation of values of complex aggregates has then to be performed. Since we are not permuting the values of a single feature, but of a whole family, we have to keep some coherence: each training example has one value for each aggregate in the family, and they should not be separated by the permutation. An example that obtains the value of a second example for a first aggregate, should not obtain the value of a third example for a second aggregate, but rather the value of the second example. In other words, for a given family of aggregates, only one permutation of examples has to be found, since a set of aggregate values for a given example should be conserved through permutation. We achieve this by permuting groups of secondary objects, i.e. the set of secondary objects related to one example will be assigned to another example. By doing this, all aggregate values are transferred from one example to another.

The family importance is then computed as described above: for each tree we obtain the error gain between before and after the permutation, and the gain average over all trees gives the final importance.

Table 5. Importance of main features and aggregation processes in urban blocks.

Feature	Score
Area	0.039
Elongation	0.003
Convexity	0.005
Density	0.157
Count	0.027
Minimum Area	0.062
Minimum Elongation	0.034
Minimum Convexity	0.028
Maximum Area	0.111
Maximum Elongation	0.054
Maximum Convexity	0.038
Average Area	0.177
Average Elongation	0.061
Average Convexity	0.039

As an example, the importances of blocks main features and buildings aggregation processes are reported in Table 5. Importances were obtained using a forest of 100 trees built using the "Random" hill-climbing heuristic to find complex aggregates.

We observe that the 3 most important features for urban blocks classification are the average area of buildings, the density of blocks, and the maximum area of buildings.

7   Conclusion and Future Work

In this paper, we presented CARAF, a relational random forest learner based on complex aggregates. The hill-climbing algorithms to explore the search space perform better than RELAGGS with Random Forests and FORF on most datasets. The basic random hill-climbing algorithms to explore the complex aggregates search space yield a considerable speed up while not suffering performance loss.

Future work will consist in exploring database technologies that are suitable for learning from relational data. Indeed, most relational algorithms have not been designed to handle big data, and there is an increasing trend towards relevant representation of relational data and the technologies, potentially NoSQL-based, fitted for relational data mining.

References

1. Anderson, G., Pfahringer, B.: Relational random forests based on random relational rules. In: Boutilier, C. (ed.) IJCAI 2009, Proceedings of the 21st International Joint Conference on Artificial Intelligence, Pasadena, California, USA, July 11–17, 2009, pp. 986–991 (2009). http://ijcai.org/papers09/Papers/IJCAI09-167.pdf
2. Blockeel, H., Raedt, L.D.: Top-down induction of first-order logical decision trees. Artif. Intell. **101**(1–2), 285–297 (1998)
3. Boullé, M.: Towards automatic feature construction for supervised classification. In: Calders, T., Esposito, F., Hüllermeier, E., Meo, R. (eds.) ECML PKDD 2014, Part I. LNCS, vol. 8724, pp. 181–196. Springer, Heidelberg (2014). http://dx.doi.org/10.1007/978-3-662-44848-9_12
4. Breiman, L.: Random forests. Mach. Learn. **45**(1), 5–32 (2001). http://dx.doi.org/10.1023/A:1010933404324
5. Charnay, C., Lachiche, N., Braud, A.: Construction of complex aggregates with random restart hill-climbing. In: Davis, J., et al. (eds.) ILP 2014. LNCS, vol. 9046, pp. 49–61. Springer, Heidelberg (2015). doi:10.1007/978-3-319-23708-4_4
6. Dietterich, T.G., Lathrop, R.H., Lozano-Pérez, T.: Solving the multiple instance problem with axis-parallel rectangles. Artif. Intell. **89**(1–2), 31–71 (1997). http://dx.doi.org/10.1016/S0004-3702(96)00034-3
7. Dzeroski, S., Schulze-Kremer, S., Heidtke, K.R., Siems, K., Wettschereck, D., Blockeel, H.: Diterpene structure elucidation from 13CNMR spectra with inductive logic programming. Appl. Artif. Intell. **12**(5), 363–383 (1998). http://dx.doi.org/10.1080/088395198117686
8. Hall, M.A., Frank, E., Holmes, G., Pfahringer, B., Reutemann, P., Witten, I.H.: The WEKA data mining software: an update. SIGKDD Explor. **11**(1), 10–18 (2009). http://doi.acm.org/10.1145/1656274.1656278
9. Krogel, M.A., Wrobel, S.: Facets of aggregation approaches to propositionalization. In: Horvath, T., Yamamoto, A. (eds.) Work-in-Progress Track at the Thirteenth International Conference on Inductive Logic Programming (ILP) (2003)
10. Puissant, A., Lachiche, N., Skupinski, G., Braud, A., Perret, J., Mas, A.: Classification et évolution des tissus urbains à partir de données vectorielles. Rev. Int. de Géomatique **21**(4), 513–532 (2011)
11. Quinlan, J.R.: C4.5: Programs for Machine Learning. Morgan Kaufmann, San Francisco (1993)
12. Srinivasan, A., Muggleton, S., Sternberg, M.J.E., King, R.D.: Theories for mutagenicity: A study in first-order and feature-based induction. Artif. Intell. **85**(1–2), 277–299 (1996). http://dx.doi.org/10.1016/0004-3702(95)00122-0
13. Van Assche, A., Vens, C., Blockeel, H., Dzeroski, S.: First order random forests: Learning relational classifiers with complex aggregates. Mach. Learn. **64**(1–3), 149–182 (2006)
14. Vens, C., Ramon, J., Blockeel, H.: Refining aggregate conditions in relational learning. In: Fürnkranz, J., Scheffer, T., Spiliopoulou, M. (eds.) PKDD 2006. LNCS (LNAI), vol. 4213, pp. 383–394. Springer, Heidelberg (2006)

Distributed Parameter Learning
for Probabilistic Ontologies

Giuseppe Cota[1]($\boxtimes$), Riccardo Zese[1], Elena Bellodi[1], Fabrizio Riguzzi[2],
and Evelina Lamma[1]

[1] Dipartimento di Ingegneria, University of Ferrara,
Via Saragat 1, 44122 Ferrara, Italy
{giuseppe.cota,riccardo.zese,elena.bellodi,evelina.lamma}@unife.it
[2] Dipartimento di Matematica e Informatica, University of Ferrara,
Via Saragat 1, 44122 Ferrara, Italy
fabrizio.riguzzi@unife.it

Abstract. Representing uncertainty in Description Logics has recently
received an increasing attention because of its potential to model real
world domains. EDGE, for "Em over bDds for description loGics paramEter learning", is an algorithm for learning the parameters of probabilistic ontologies from data. However, the computational cost of this
algorithm is significant since it may take hours to complete an execution.
In this paper we present EDGE^{MR}, a distributed version of EDGE that
exploits the MapReduce strategy by means of the Message Passing Interface. Experiments on various domains show that EDGE^{MR} significantly
reduces EDGE running time.

Keywords: Probabilistic description logics · Parameter learning ·
MapReduce · Message Passing Interface

1 Introduction

Representing uncertain information is becoming crucial to model real world
domains. The ability to describe and reason with probabilistic knowledge bases
is a well-known topic in the field of Description Logics (DLs). In order to
model domains with complex and uncertain relationships, several approaches
have been proposed that combine logic and probability. The distribution
semantics [22] from Logic Programming is one of them. In [3,14,21,27] the
authors proposed an approach for the integration of probabilistic information in
DLs called DISPONTE (for "DIstribution Semantics for Probabilistic ONTologiEs"), which adapts the distribution semantics for probabilistic logic programs
to DLs.

EDGE [16], for "Em over bDds for description loGics paramEter learning", is an algorithm for learning the parameters of probabilistic DLs following
DISPONTE. EDGE was tested on various datasets and was able to find good
solutions. However, the algorithm may take hours on datasets whose size is of

© Springer International Publishing Switzerland 2016
K. Inoue et al. (Eds.): ILP 2015, LNAI 9575, pp. 30–45, 2016.
DOI: 10.1007/978-3-319-40566-7_3

the order of MBs. In order to efficiently manage larger datasets in the era of Big Data, it is of foremost importance to develop approaches for reducing the learning time. One solution is to distribute the algorithm using modern computing infrastructures such as clusters and clouds.

Here we present EDGE^{MR}, a MapReduce version of EDGE. MapReduce [8] is a model for processing data in parallel on a cluster. In this model the work is distributed among mapper and reducer workers. The mappers take the input data and return a set of (key, value) pairs. These sets are then grouped according to the key into couples (key, set_of_values) and the reducers aggregate the values obtaining a set of (key, aggregated_value) couples that represents the output of the task.

Various MapReduce frameworks are available, such as Hadoop[1]. However we chose not to use any framework and to implement a much simpler MapReduce approach for EDGE^{MR} based on the Message Passing Interface (MPI). The reason is that we adopted a modified MapReduce approach where the map and reduce workers are not purely functional in order to better adapt to the task at hand.

A performance evaluation of EDGE^{MR} is provided through a set of experiments on various datasets using 1, 3, 5, 9 and 17 computing nodes. The results show that EDGE^{MR} effectively reduces EDGE running time, with good speedups.

The paper is structured as follows. Section 2 introduces Description Logics while Sect. 3 summarizes DISPONTE and an inference system for probabilistic DLs. Section 4 briefly describes EDGE and Sect. 5 presents EDGE^{MR}. Section 6 shows the results of the experiments for evaluating EDGE^{MR}, while Sect. 7 discusses related works. Section 8 draws conclusions.

2 Description Logics

DLs are a family of logic based knowledge representation formalisms which are of particular interest for representing ontologies in the Semantic Web. For a good introduction to DLs we refer to [2].

While DLs are a fragment of first order logic, they are usually represented using a syntax based on concepts and roles. A concept corresponds to a set of individuals of the domain while a role corresponds to a set of couples of individuals of the domain. The proposed algorithm can deal with $\mathcal{SROIQ}(\mathbf{D})$ DLs which we describe in the following.

We use $\mathbf{A}$, $\mathbf{R}$ and $\mathbf{I}$ to indicate *atomic concepts*, *atomic roles* and *individuals*, respectively. A *role* could be an atomic role $R \in \mathbf{R}$, the inverse R^- of an atomic role $R \in \mathbf{R}$ or a complex role $R \circ S$. We use $\mathbf{R}^-$ to denote the set of all inverses of roles in $\mathbf{R}$. Each $A \in \mathbf{A}$, $\bot$ and $\top$ are concepts and if $a \in \mathbf{I}$, then $\{a\}$ is a concept called *nominal*. If C, C_1 and C_2 are concepts and $R \in \mathbf{R} \cup \mathbf{R}^-$, then $(C_1 \sqcap C_2)$, $(C_1 \sqcup C_2)$ and $\neg C$ are concepts, as well as $\exists R.C$, $\forall R.C$, $\geq nR.C$ and $\leq nR.C$ for an integer $n \geq 0$.

---

[1] http://hadoop.apache.org/.

A *knowledge base* (KB) $\mathcal{K} = (\mathcal{T}, \mathcal{R}, \mathcal{A})$ consists of a TBox $\mathcal{T}$, an RBox $\mathcal{R}$ and an ABox $\mathcal{A}$. An RBox $\mathcal{R}$ is a finite set of *transitivity axioms* $Trans(R)$, *role inclusion axioms* $R \sqsubseteq S$ and *role chain axioms* $R \circ P \sqsubseteq S$, where $R, P, S \in \mathbf{R} \cup \mathbf{R}^-$. A *TBox* $\mathcal{T}$ is a finite set of *concept inclusion axioms* $C \sqsubseteq D$, where C and D are concepts. An *ABox* $\mathcal{A}$ is a finite set of *concept membership axioms* $a : C$ and *role membership axioms* $(a, b) : R$, where C is a concept, $R \in \mathbf{R}$ and $a, b \in \mathbf{I}$.

A KB is usually assigned a semantics using interpretations of the form $\mathcal{I} = (\Delta^{\mathcal{I}}, \cdot^{\mathcal{I}})$, where $\Delta^{\mathcal{I}}$ is a non-empty *domain* and $\cdot^{\mathcal{I}}$ is the *interpretation function* that assigns an element in $\Delta^{\mathcal{I}}$ to each individual a, a subset of $\Delta^{\mathcal{I}}$ to each concept C and a subset of $\Delta^{\mathcal{I}} \times \Delta^{\mathcal{I}}$ to each role R. The mapping $\cdot^{\mathcal{I}}$ is extended to complex concepts as follows (where $R^{\mathcal{I}}(x, C) = \{y | \langle x, y \rangle \in R^{\mathcal{I}}, y \in C^{\mathcal{I}}\}$ and $\#X$ denotes the cardinality of the set X):

$$\top^{\mathcal{I}} = \Delta^{\mathcal{I}}$$
$$\bot^{\mathcal{I}} = \emptyset$$
$$\{a\}^{\mathcal{I}} = \{a^{\mathcal{I}}\}$$
$$(\neg C)^{\mathcal{I}} = \Delta^{\mathcal{I}} \setminus C^{\mathcal{I}}$$
$$(C_1 \sqcup C_2)^{\mathcal{I}} = C_1^{\mathcal{I}} \cup C_2^{\mathcal{I}}$$
$$(C_1 \sqcap C_2)^{\mathcal{I}} = C_1^{\mathcal{I}} \cap C_2^{\mathcal{I}}$$
$$(\exists R.C)^{\mathcal{I}} = \{x \in \Delta^{\mathcal{I}} | R^{\mathcal{I}}(x) \cap C^{\mathcal{I}} \neq \emptyset\}$$
$$(\forall R.C)^{\mathcal{I}} = \{x \in \Delta^{\mathcal{I}} | R^{\mathcal{I}}(x) \subseteq C^{\mathcal{I}}\}$$
$$(\geq nR.C)^{\mathcal{I}} = \{x \in \Delta^{\mathcal{I}} | \#R^{\mathcal{I}}(x, C) \geq n\}$$
$$(\leq nR.C)^{\mathcal{I}} = \{x \in \Delta^{\mathcal{I}} | \#R^{\mathcal{I}}(x, C) \leq n\}$$
$$(R^-)^{\mathcal{I}} = \{(y, x) | (x, y) \in R^{\mathcal{I}}\}$$
$$(R_1 \circ ... \circ R_n)^{\mathcal{I}} = R_1^{\mathcal{I}} \circ ... \circ R_n^{\mathcal{I}}$$

A query over a KB is an axiom for which we want to test the entailment from the KB.

3   Semantics and Reasoning in Probabilistic DLs

DISPONTE [3] applies the distribution semantics to probabilistic ontologies [22]. In DISPONTE a *probabilistic knowledge base* $\mathcal{K}$ is a set of certain and probabilistic axioms. *Certain axioms* are regular DL axioms. *Probabilistic axioms* take the form $p :: E$, where p is a real number in $[0, 1]$ and E is a DL axiom.

The idea of DISPONTE is to associate independent Boolean random variables with the probabilistic axioms. By assigning values to every random variable we obtain a *world*, i.e. the union of the probabilistic axioms whose random variable takes on value 1 and the set of certain axioms.

The probability p can be interpreted as an *epistemic probability*, i.e., as the degree of our belief in axiom E. For example, a probabilistic concept membership axiom $p :: a : C$ means that we have degree of belief p in $C(a)$. The statement that Tweety flies with probability 0.9 can be expressed as $0.9 :: tweety : Flies$.

Let us now give the formal definition of DISPONTE. An *atomic choice* is a pair (E_i, k) where E_i is the ith probabilistic axiom and $k \in \{0, 1\}$. k indicates whether E_i is chosen to be included in a world ($k = 1$) or not ($k = 0$).

A *composite choice* κ is a consistent set of atomic choices, i.e. $(E_i, k) \in \kappa, (E_i, m) \in \kappa \Rightarrow k = m$ (only one decision for each axiom). The probability of a composite choice κ is $P(\kappa) = \prod_{(E_i,1)\in\kappa} p_i \prod_{(E_i,0)\in\kappa}(1-p_i)$, where p_i is the probability associated with axiom E_i. A *selection* σ is a composite choice that contains an atomic choice (E_i, k) for every probabilistic axiom of the theory. A selection σ identifies a theory w_σ called a *world* in this way: $w_\sigma = \{E_i|(E_i,1) \in \sigma\}$. Let us indicate with $\mathcal{S}_{\mathcal{K}}$ the set of all selections and with $\mathcal{W}_{\mathcal{K}}$ the set of all worlds. The probability of a world w_σ is $P(w_\sigma) = P(\sigma) = \prod_{(E_i,1)\in\sigma} p_i \prod_{(E_i,0)\in\sigma}(1-p_i)$. $P(w_\sigma)$ is a probability distribution over worlds, i.e. $\sum_{w\in\mathcal{W}_{\mathcal{K}}} P(w) = 1$. We can now assign probabilities to queries. Given a world w, the probability of a query Q is defined as $P(Q|w) = 1$ if $w \models Q$ and 0 otherwise. The probability of a query can be defined by marginalizing the joint probability of the query and the worlds:

$$P(Q) = \sum_{w\in\mathcal{W}_{\mathcal{K}}} P(Q,w) = \sum_{w\in\mathcal{W}_{\mathcal{K}}} P(Q|w)P(w) = \sum_{w\in\mathcal{W}_{\mathcal{K}}:w\models Q} P(w) \qquad (1)$$

The system BUNDLE [15, 17, 21] computes the probability of a query w.r.t. KBs that follow DISPONTE by first computing all the explanations for the query and then building a Binary Decision Diagram (BDD) that represents them. An explanations κ for a query Q is a composite choice that identifies a set of worlds which all entail Q. Given the set K of all explanations for a query Q, we can define the Disjunctive Normal Form (DNF) Boolean formula f_K as $f_K(\mathbf{X}) = \bigvee_{\kappa\in K} \bigwedge_{(E_i,1)} X_i \bigwedge_{(E_i,0)} \overline{X_i}$. The variables $\mathbf{X} = \{X_i|(E_i,k) \in \kappa, \kappa \in K\}$ are independent Boolean random variables with $P(X_i = 1) = p_i$ and the probability that $f_K(\mathbf{X})$ takes value 1 gives the probability of Q. A BDD for a function of Boolean variables is a rooted graph that has one level for each Boolean variable. A node n has two children: one corresponding to the 1 value of the variable associated with the level of n and one corresponding to the 0 value of the variable. When drawing BDDs, the 0-branch is distinguished from the 1-branch by drawing it with a dashed line. The leaves store either 0 or 1.

BUNDLE finds the set K of all explanations for a query Q by means of the Pellet reasoner [23]. Then it builds a BDD representing f_K, from which the probability $P(f_K = 1)$, and thus the probability of Q, can be computed with a dynamic programming algorithm in polynomial time in the size of the diagram [7].

Example 1. Let us consider the following knowledge base, inspired by the ontology `people+pets` proposed in [13]:

$$\exists hasAnimal.Pet \sqsubseteq NatureLover$$
$$(kevin, fluffy) : hasAnimal$$
$$(kevin, tom) : hasAnimal$$

(E_1) $0.4 :: fluffy : Cat$

(E_2) $0.3 :: tom : Cat$

(E_3) $0.6 :: Cat \sqsubseteq Pet$

Individuals that own an animal which is a pet are nature lovers and *kevin* owns the animals *fluffy* and *tom*. *fluffy* and *tom* are cats and cats are pets with the specified probabilities. This KB has eight worlds and the query axiom $Q = kevin : Nature Lover$ is true in three of them, corresponding to the following selections: $\{(E_1, 1), (E_2, 0), (E_3, 1)\}, \{(E_1, 0), (E_2, 1), (E_3, 1)\}, \{(E_1, 1), (E_2, 1), (E_3, 1)\}$. The probability is $P(Q) = 0.4 \cdot 0.7 \cdot 0.6 + 0.6 \cdot 0.3 \cdot 0.6 + 0.4 \cdot 0.3 \cdot 0.6 = 0.348$. If we associate the random variables X_1 with axiom E_1, X_2 with E_2 and X_3 with E_3 the BDD representing the set of explanations is shown in Fig. 1.

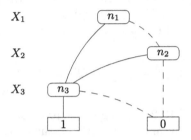

Fig. 1. BDD representing the set of explanations for the query of Example 1.

4   Parameter Learning for Probabilistic DLs

EDGE [16] adapts the algorithm EMBLEM[2] [4], developed for learning the parameters of probabilistic logic programs, to the case of probabilistic DLs under DISPONTE. Inspired by [10], it performs an Expectation-Maximization cycle over BDDs.

EDGE performs supervised parameter learning. It takes as input a DL KB and a number of positive and negative examples that represent the queries in the form of concept assertions, i.e. in the form $a : C$ for an individual a and a class C. Positive examples represent information that we regard as true and for which we would like to get high probability while negative examples represent information that we regard as false and for which we would like to get low probability.

First, EDGE generates, for each example (query), the BDD encoding its explanations using BUNDLE. Then, EDGE starts the EM cycle in which the steps of Expectation and Maximization are iterated until a local maximum of the log-likelihood (LL) of the examples is reached. The LL of the examples is guaranteed to increase at each iteration. EDGE stops when the difference between the LL of the current iteration and that of the previous one drops below a threshold ϵ or when this difference is below a fraction δ of the previous LL. Finally, EDGE returns the reached LL and the parameters p_i of the probabilistic axioms. EDGE's main procedure is illustrated in Algorithm 1.

---

[2] EMBLEM is included in the web application http://cplint.lamping.unife.it/ [20].

Algorithm 1. Function EDGE

```
function EDGE(𝒦, P_E, N_E, ε, δ)              ▷ P_E, N_E: positive and negative examples
    Build BDDs                                ▷ performed by BUNDLE
    LL = -∞
    repeat
        LL₀ = LL
        LL = EXPECTATION(BDDs)
        MAXIMIZATION
    until LL - LL₀ < ε ∨ LL - LL₀ < -LL₀ · δ
    return LL, pᵢ for all probabilistic axioms
end function
```

Function EXPECTATION (shown in Algorithm 2) takes as input a list of BDDs, one for each example Q, and computes the expectations $\mathbf{E}[c_{i0}|Q]$ and $\mathbf{E}[c_{i1}|Q]$ for all axioms E_i directly over the BDDs. c_{ix} represents the number of times a Boolean random variable X_i takes value x for $x \in \{0,1\}$ and $\mathbf{E}[c_{ix}|Q] = P(X_i = x|Q)$. Then it sums up the contributions of all examples: $\mathbf{E}[c_{ix}] = \sum_Q \mathbf{E}[c_{ix}|Q]$.

Algorithm 2. Function EXPECTATION

```
function EXPECTATION(BDDs)
    LL = 0
    for all i ∈ Axioms do
        E[c_{i0}] = E[c_{i1}] = 0
    end for
    for all BDD ∈ BDDs do
        for all i ∈ Axioms do
            η⁰(i) = η¹(i) = 0
        end for
        for all variables X do
            ς(X) = 0
        end for
        GETFORWARD(root(BDD))
        Prob=GETBACKWARD(root(BDD))
        T = 0
        for l = 1 to levels(BDD) do
            Let X_i be the variable associated with level l
            T = T + ς(X_i)
            η⁰(i) = η⁰(i) + T · (1 - p_i)
            η¹(i) = η¹(i) + T · p_i
        end for
        for all i ∈ Axioms do
            E[c_{i0}] = E[c_{i0}] + η⁰(i)/Prob
            E[c_{i1}] = E[c_{i1}] + η¹(i)/Prob
        end for
        LL = LL + log(Prob)
    end for
    return LL
end function
```

In turn, $P(X_i = x|Q)$ is given by $\frac{P(X_i=x,Q)}{P(Q)}$. In Algorithm 2 we use $\eta^x(i)$ to indicate $\mathbf{E}[c_{ix}]$. EXPECTATION first calls procedures GETFORWARD and GET-BACKWARD that compute the forward and the backward probability of nodes and $\eta^x(i)$ for non-deleted paths only. These are the paths that have not been deleted when building the BDDs. Forward and backward probabilities in each node represent the probability mass of paths from the root to the node and that of the paths from the node to the leaves respectively. The expression

$$P(X_i = x, Q) = \sum_{n \in N(Q), v(n) = X_i} F(n)B(child_x(n))\pi_{ix}$$

represents the total probability mass of each path passing through the nodes associated with X_i and going down its x-branch, with $N(Q)$ the set of BDD nodes for query Q, $v(n)$ the variable associated with node n, $\pi_{i1} = p_i$, $\pi_{i0} = 1 - p_i$, $F(n)$ the forward probability of n, $B(n)$ the backward probability of n.

Computing the two probabilities in the nodes requires two traversals of the graph, so its cost is linear in the number of nodes. Procedure GETFORWARD computes the forward probabilities for every node. It traverses the diagram one level at a time starting from the root level, where $F(root) = 1$, and for each node n computes its contribution to the forward probabilities of its children. Then the forward probabilities of both children are updated. Function GET-BACKWARD computes backward probabilities of nodes by traversing recursively the tree from the leaves to the root. It returns the backward probability of the root corresponding to the probability of the query $P(Q)$, indicated with $Prob$ in Algorithm 2.

When the calls of GETBACKWARD for both children of a node n return, we have the $e^x(n)$ and $\eta^x(i)$ values for non-deleted paths only. An array ς is used to store the contributions of the deleted paths that is then added to $\eta^x(i)$. See [16] for more details.

Expectations are updated for all axioms and finally the log-likelihood of the current example is added to the overall LL.

Function MAXIMIZATION computes the parameters' values for the next EM iteration by relative frequency. Note that the $\eta^x(i)$ values can be stored in a bi-dimensional array $eta[a, 2]$ where a is the number of axioms.

EDGE is written in Java for portability and interfacing with Pellet. For further information about EDGE please refer to [16].

5   Distributed Parameter Learning for Probabilistic DLs

The aim of the present work is to develop a parallel version of EDGE that exploits the MapReduce strategy. We called it EDGE^{MR} (see Algorithm 3).

5.1   Architecture

Like most MapReduce frameworks, EDGE^{MR} architecture follows a master-slave model. The communication between the master and the slaves adopts the Message Passing Interface (MPI), in particular we used the OpenMPI[3] library which provides a Java interface to the native library. The processes of EDGE^{MR} are not purely functional, as required by standard MapReduce frameworks such as

---

[3] http://www.open-mpi.org/.

Hadoop, because they have to retain in main memory the BDDs during the whole execution. This forced us to develop a parallelization strategy exploiting MPI.

EDGE^{MR} can be split into three phases: *Initialization, Query resolution* and *Expectation-Maximization*. All these operations are executed in parallel and synchronized by the master.

Initialization. During this phase data is replicated and a process is created on each machine. Then each process parses its copy of the probabilistic knowledge base and stores it in main memory. The master, in addition, parses the files containing the positive and negative examples (the queries).

Query resolution. The master divides the set of queries into subsets and distributes them among the workers. Each worker generates its private subset of BDDs and keeps them in memory for the whole execution. Two different scheduling techniques can be applied for this operation. See Subsect. 5.3 for details.

Expectation-Maximization. After all the nodes have built the BDDs for their queries, EDGE^{MR} starts the Expectation-Maximization cycle. During the Expectation step all the workers traverse their BDDs and calculate their local *eta* array. Then the master gathers all the *eta*'s from the workers and aggregates them by summing the arrays component-wise. Then it calls the Maximization procedure in which it updates the parameters and sends them to the slaves. The cycle is repeated until one of the stopping criteria is satisfied.

5.2 MapReduce View

Since EDGE^{MR} is based on MapReduce, it can be split into three phases: *Initialization, Map* and *Reduce*.

Initialization. Described in Subsect. 5.1.

Map. This phase can be seen as a function that returns a set of *(key, value)* pairs, where *key* is an example identifier and *value* is the array *eta*.
  - *Query resolution*: each worker resolves its chunks of queries and builds its private set of BDDs. Two different scheduling techniques can be applied for this operation. See Subsect. 5.3 for details.
  - *Expectation Step*: each worker calculates its local *eta*.

Reduce. This phase is performed by the master (also referred to as the "reducer") and it can be seen as a function that returns pairs (i, p_i), where i is an axiom identifier and p_i is its probability.
  - *Maximization Step*: the master gathers all the *eta* arrays from the workers, aggregates them by summing component wise, performs the Maximization step and sends the newly updated parameters to the slaves.

The Map and Reduce phases implement the Expectation and Maximization functions respectively, hence they are repeated until a local maximum is reached. It is important to notice that the Query Resolution step in the Map phase is executed only once because the workers keep in memory the generated BDDs for the whole execution of the EM cycle; what changes among iterations are the random variables' parameters.

5.3   Scheduling Techniques

In a distributed context the scheduling strategy influences significantly the performances. We evaluated two scheduling strategies, *single-step scheduling* and *dynamic scheduling*, during the generation of the BDDs for the queries, while the initialization and the EM phases are independent of the chosen scheduling method.

Single-step Scheduling if N is the number of the slaves, the master divides the total number of queries envenly into $N + 1$ chunks, i.e. the number of slaves plus the master. Then the master starts $N+1$ threads, one building the BDD for its queries while the others sending the other chunks to the corresponding slaves. After the master has terminated dealing with its queries, it waits for the results from the slaves. When the slowest slave returns its results to the master, EDGE^{MR} proceeds to the EM cycle. Figure 2(a) shows an example of single-step scheduling with two slaves.

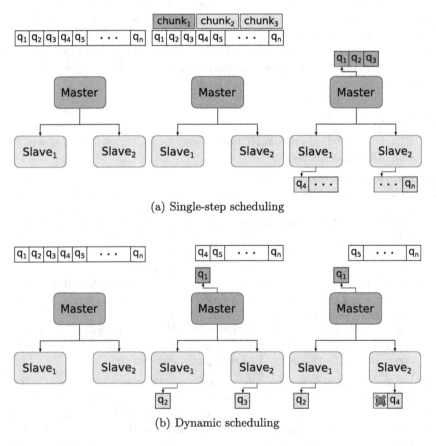

(a) Single-step scheduling

(b) Dynamic scheduling

Fig. 2. Scheduling techniques of EDGE^{MR}.

Algorithm 3. Function EDGE^{MR}

1: **function** EDGE^{MR} $(\mathcal{K}, P_E, N_E, S, \epsilon, \delta)$ ▷ P_E, N_E: pos. and neg. examples, S: scheduling method
2:    Read knowledge base $\mathcal{K}$
3:    **if** MASTER **then**
4:        Identify examples E
5:        **if** $S ==$ dynamic **then** ▷ dynamic scheduling
6:            Send a chunk of examples E_j to each slave
7:            Start thread listener ▷ Thread for answering query requests from slaves
8:            $c = m - 1$ ▷ c counts the computed examples
9:            **while** $c < |E|$ **do**
10:               $c = c + 1$
11:               Build BDD_c for example e_c
12:           **end while**
13:       **else** ▷ single-step scheduling
14:           Split examples E evenly into n subsets $E_1, \ldots, E_n$
15:           Send E_m to each worker $m, 2 \leq m \leq n$
16:           Build $BDDs_1$ for examples E_1
17:       **end if**
18:       $LL = -\infty$
19:       **repeat**
20:           $LL_0 = LL$
21:           Send the parameters p_i to each worker $m, 2 \leq m \leq n$
22:           $LL = $ EXPECTATION$(BDDs_1)$
23:           Collect LL_m and the expectations from each worker $m, 2 \leq m \leq n$
24:           Update LL and the expectations
25:           MAXIMIZATION
26:       **until** $LL - LL_0 < \epsilon \vee LL - LL_0 < -LL \cdot \delta$
27:       Send STOP signal to all slaves
28:       **return** LL, p_i for all i
29:   **else** ▷ the j-th slave
30:       **if** $S ==$ dynamic **then** ▷ dynamic scheduling
31:           **while** $c < |E|$ **do**
32:               Receive E_j from master
33:               Build $BDDs_j$ for the chunk of examples E_j
34:               Request another chunk of examples to the master
35:           **end while**
36:       **else** ▷ single-step scheduling
37:           Receive E_j from master
38:           Build $BDDs_j$ for examples E_j
39:       **end if**
40:       **repeat**
41:           Receive the parameters p_i from master
42:           $LL_j = $ EXPECTATION$(BDDs_j)$
43:           Send LL_j and the expectations to master
44:       **until** Receive STOP signal from master
45:   **end if**
46: **end function**

Dynamic Scheduling is more flexible and adaptive than single-step scheduling. Handling each query chunk may require a different amount of time. Therefore, with single-step scheduling, it could happen that a slave takes much more time than another one to deal with its chunk of queries. This may cause the master and some slaves to wait. Dynamic scheduling mitigates this issue. The user can establish a chunk dimension, i.e. the number of examples in each chunk. At first, each machine is assigned a chunk of queries in order. When it finishes handling the chunk, it takes the following chunk. So if the master ends handling its chunk, it just picks the next one, while if a slave ends handling its chunk, it asks the master for another one.

During this phase the master runs a listener thread that waits for slaves' requests of new chunks. For each request, the listener starts a new thread that sends a chunk to the requesting slave (to improve the performances this is done through a thread pool). When all the BDDs for the queries are built, EDGE^{MR} starts the EM cycle. An example of dynamic scheduling with two slaves and a chunk dimension of one example is displayed in Fig. 2(b).

5.4 Overall EDGE^{MR}

In light of the above, we analyze now EDGE^{MR}'s main procedure, shown in Algorithm 3.

After the *initialization* phase, EDGE^{MR} enters in the *query resolution* phase, where the master sends the examples to the slaves and builds its BDDs [lines 4–17]. Here, in particular, if dynamic scheduling is chosen, the master initializes a thread listener (line 7) which sends a chunk of examples to the slaves at every request it receives, while if single-step scheduling is chosen it simply divides the examples evenly and sends them once for all to the slaves. Meanwhile, the slaves, following the chosen scheduling type, receive the examples and build the corresponding BDDs (lines 30–39). In the dynamic scheduling setting the slaves make a request to the master for another chunk of queries every time they finish to compute the current chunk (line 34). After that, EDGE^{MR} enters in the *Expectation-Maximization* phase, first of all the master sends the probability values p_i to the slaves (line 21) which receive and store them (line 41). Now, the EXPECTATION procedure (Algorithm 2) can be executed by all the workers (lines 22 and 42). Finally, during the maximization phase the master collects all the values, executes the MAXIMIZATION procedure and checks whether a new round of EM must be performed (lines 23–27), while the slaves only wait for a signal from master which indicates whether to execute either EXPECTATION or stop.

6 Experiments

In order to evaluate the performances of EDGE^{MR}, four datasets were selected:

- **Mutagenesis**[4] [25], contains information about a number of aromatic and heteroaromatic nitro drugs, including their chemical structures in terms of atoms, bonds and a number of molecular substructures.
- **Carcinogenesis**[5] [24], which describes the carcinogenicity of more than 300 chemical compounds.
- an extract of **DBPedia**[6] [11], a knowledge base obtained by extracting the structured data from Wikipedia.

---

[4] http://www.doc.ic.ac.uk/~shm/mutagenesis.html.

[5] http://dl-learner.org/wiki/Carcinogenesis.

[6] http://dbpedia.org/.

– `education.data.gov.uk`[7], which contains information about school institutions in the United Kingdom.

The last three datasets are the same as in [19]. All experiments have been performed on a cluster of 64-bit Linux machines with 2 GB (max) memory allotted to Java per node. Each node of this cluster has 8-cores Intel Haswell 2.40 GHz CPUs.

For the generation of positive and negative examples, we randomly chose a set of individuals from the dataset. Then, for each extracted individual a, we sampled three named classes: A and B were selected among the named classes to which a explicitly belongs, while C was taken from the named classes to which a does not explicitly belong but that exhibits at least one explanation for the query $a : C$. The axiom $a : A$ was added to the KB, while $a : B$ was considered as a positive example and $a : C$ as a negative example. Then both the positive and the negative examples were split in five equally sized subsets and we performed five-fold cross-validation for each dataset and for each number of workers. Information about the datasets and training examples is shown in Table 1. We performed the experiments with 1, 3, 5, 9 and 17 nodes, where the execution with 1 node corresponds to the execution of EDGE. Furthermore, we used both single-step and dynamic scheduling in order to evaluate the two scheduling approaches. It is important to point out that the quality of the learning is independent of the type of scheduling and of the number of nodes, i.e. the parameters found with 1 node are the same as those found with n nodes. Table 2 shows the running time in seconds for parameter learning on the four datasets with the different configurations. Figure 3 shows the speedup obtained as a function of the number of machines (nodes). The speedup is the ratio of the running time of 1 worker to the running time of n workers. We can note that the speedup is significant even if it is sublinear, showing that a certain amount of overhead (the resources, and therefore the time, spent for the MPI communications) is present. The dynamic scheduling technique has generally better performance than single-step scheduling.

Table 1. Characteristics of the datasets used for evaluation.

Dataset	# of all axioms	# of probabilistic axioms	# of pos. examples	# of neg. examples	Fold size (MiB)
Carcinogenesis	74409	186	103	154	18.64
DBpedia	5380	1379	181	174	0.98
`education.data.gov.uk`	5467	217	961	966	1.03
Mutagenesis	48354	92	500	500	6.01

---

[7] http://education.data.gov.uk.

Table 2. Comparison between EDGE and EDGE^{MR} in terms of running time (in seconds) for parameter learning.

Dataset	EDGE	EDGE^{MR}							
		Dynamic				Single-step			
		3	5	9	17	3	5	9	17
Carcinogenesis	847	441.8	241	147.2	94.2	384	268.4	179.2	117.8
DBpedia	1552	1259.8	634	364.6	215.2	1155.6	723.8	452.6	372.6
education.data.gov.uk	6924.2	3878.2	2157.2	1086	623.2	3611.6	2289.6	1331.6	749.4
Mutagenesis	1439.4	635.8	399.8	223.2	130.4	578.2	359.2	230	124.6

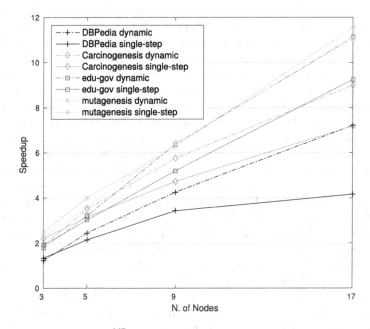

Fig. 3. Speedup of EDGE^{MR} relative to EDGE with single-step and dynamic schedulings.

7   Related Work

The pervasiveness of Internet, the availability of sensor data, the dramatically increased storage and computational capabilities provide the opportunity to gather huge sets of data. This large amount of information is known by the name of *Big Data*. The ability to process and perform learning and inference over massive data is one of the major challenges of the current decade. Big data is strictly intertwined with the availability of scalable and distributed algorithms.

In [1] the authors developed a method to parallelize inference and training on a probabilistic relational model. They show that (loopy) belief propagation can be lifted and that lifting is MapReduce-able. In addition they show that the MapReduce approach improves performances significantly. For parameter learning, they propose an approximate method which is MapReduce-able as well. In order to train large models, they shatter the factor graph into smaller pieces which can be elaborated locally in a distributed fashion by exploiting the MapReduce approach.

Other specific distributed implementations have been developed for various machine learning methods, such as support vector machines [26], robust regression [12] and extreme learning machines [9].

8 Conclusions

EDGE is an algorithm for learning the parameters of DISPONTE probabilistic knowledge bases. In this paper we presented EDGE^{MR}, a distributed version of EDGE based on the MapReduce approach.

We performed experiments over four datasets with an increasing number of nodes. The results show that parallelization significantly reduces the execution time, even if with a sublinear trend due to overhead.

We are currently working on a way to distribute structure learning of DISPONTE probabilistic knowledge bases. In particular, we aim at developing a MapReduce version of LEAP [19], as outlined in [5,6]. Moreover, we plan to investigate the possibility of parallelizing and distributing also the inference. We also plan to develop a Web interface for EDGE^{MR} and integrate it in TRILL on SWISH, available at http://trill.lamping.unife.it/, possibly reimplementing it in Prolog, using the technology of [18,28].

References

1. Ahmadi, B., Kersting, K., Mladenov, M., Natarajan, S.: Exploiting symmetries for scaling loopy belief propagation and relational training. Mach. Learn. **92**(1), 91–132 (2013)
2. Baader, F., Calvanese, D., McGuinness, D.L., Nardi, D., Patel-Schneider, P.F. (eds.): The Description Logic Handbook: Theory, Implementation, and Applications. Cambridge University Press, New York (2003)
3. Bellodi, E., Lamma, E., Riguzzi, F., Albani, S.: A distribution semantics for probabilistic ontologies. In: Proceedings of the 7th International Workshop on Uncertainty Reasoning for the Semantic Web. CEUR Workshop Proceedings, vol. 778, pp. 75–86. Sun SITE Central Europe (2011)
4. Bellodi, E., Riguzzi, F.: Expectation Maximization over Binary DecisionDiagrams for probabilistic logic programs. Intell. Data Anal. **17**(2), 343–363 (2013)
5. Cota, G., Zese, R., Bellodi, E., Lamma, E., Riguzzi, F.: Learning probabilistic ontologies with distributed parameter learning. In: Bellodi, E., Bonfietti, A. (eds.) Proceedings of the Doctoral Consortium (DC) co-located with the 14th Conference of the Italian Association for Artificial Intelligence (AI*IA 2015). CEUR Workshop Proceedings, vol. 1485, pp. 7–12. Sun SITE Central Europe, Aachen (2015)
6. Cota, G., Zese, R., Bellodi, E., Lamma, E., Riguzzi, F.: Structure learning with distributed parameter learning for probabilistic ontologies. In: Hollmen, J., Papapetrou, P. (eds.) Doctoral Consortium of the European Conference on Machine Learning and Principles and Practice of Knowledge Discovery in Databases (ECMLPKDD 2015), pp. 75–84 (2015). http://urn.fi/URN:ISBN: 978-952-60-6443-7

7. De Raedt, L., Kimmig, A., Toivonen, H.: ProbLog: A probabilistic Prolog and its application in link discovery. In: Proceedings of the Twentieth International Joint Conference on Artificial Intelligence, Hyderabad, India (IJCAI-05). vol. 7, pp. 2462–2467. AAAI Press (2007)

8. Dean, J., Ghemawat, S.: MapReduce: simplified data processing on large clusters. Commun. ACM **51**(1), 107–113 (2008)

9. He, Q., Shang, T., Zhuang, F., Shi, Z.: Parallel extreme learning machine for regression based on mapreduce. Neurocomputing **102**, 52–58 (2013)

10. Ishihata, M., Kameya, Y., Sato, T., Minato, S.: Propositionalizing the EM algorithm by BDDs. In: Late Breaking Papers of the 18th International Conference on Inductive Logic Programming (ILP 2008), pp. 44–49 (2008)

11. Lehmann, J., Isele, R., Jakob, M., Jentzsch, A., Kontokostas, D., Mendes, P.N., Hellmann, S., Morsey, M., van Kleef, P., Auer, S., Bizer, C.: DBpedia - a large-scale, multilingual knowledge base extracted from wikipedia. Semantic Web **6**(2), 167–195 (2015)

12. Meng, X., Mahoney, M.: Robust regression on mapreduce. In: Proceedings of the 30th International Conference on Machine Learning, pp. 888–896. JMLR (2013)

13. Patel-Schneider, P., F., Horrocks, I., Bechhofer, S.: Tutorial on OWL (2003)

14. Riguzzi, F., Bellodi, E., Lamma, E., Zese, R.: Epistemic and statistical probabilistic ontologies. In: Proceedings of the 8th International Workshop on Uncertainty Reasoning for the Semantic Web. CEUR Workshop Proceedings, vol. 900, pp. 3–14. Sun SITE Central Europe (2012)

15. Riguzzi, F., Bellodi, E., Lamma, E., Zese, R.: Computing instantiated explanations in OWL DL. In: Baldoni, M., Baroglio, C., Boella, G., Micalizio, R. (eds.) AI*IA 2013. LNCS, vol. 8249, pp. 397–408. Springer, Heidelberg (2013)

16. Riguzzi, F., Bellodi, E., Lamma, E., Zese, R.: Parameter learning for probabilistic ontologies. In: Faber, W., Lembo, D. (eds.) RR 2013. LNCS, vol. 7994, pp. 265–270. Springer, Heidelberg (2013)

17. Riguzzi, F., Bellodi, E., Lamma, E., Zese, R.: Probabilistic description logics under the distribution semantics. Seman. Web **6**(5), 447–501 (2015)

18. Riguzzi, F., Bellodi, E., Lamma, E., Zese, R.: Reasoning with probabilistic ontologies. In: Yang, Q., Wooldridge, M. (eds.) Proceedings of the Twenty-Fourth International Joint Conference on Artificial Intelligence, Buenos Aires, Argentina, pp. 4310–4316. AAAI Press/International Joint Conferences on Artificial Intelligence, Palo Alto (2015)

19. Riguzzi, F., Bellodi, E., Lamma, E., Zese, R., Cota, G.: Learning probabilistic description logics. In: Bobillo, F., et al. (eds.) URSW III. LNCS, vol. 8816, pp. 63–78. Springer, Heidelberg (2014)

20. Riguzzi, F., Bellodi, E., Lamma, E., Zese, R., Cota, G.: Probabilistic logic programming on the web. Software Pract. and Exper (2016, to appear). http://ds.ing.unife.it/~friguzzi/Papers/RigBelLam-SPE16.pdf

21. Riguzzi, F., Bellodi, E., Lamma, E., Zese, R.: BUNDLE: a reasoner for probabilistic ontologies. In: Faber, W., Lembo, D. (eds.) RR 2013. LNCS, vol. 7994, pp. 183–197. Springer, Heidelberg (2013)

22. Sato, T.: A statistical learning method for logic programs with distribution semantics. In: Sterling, L. (ed.) Proceedings of the Twelfth International Conference on Logic Programming, Tokyo, Japan, pp. 715–729. MIT Press (1995)

23. Sirin, E., Parsia, B., Cuenca-Grau, B., Kalyanpur, A., Katz, Y.: Pellet: a practical OWL-DL reasoner. J. Web Semant. **5**(2), 51–53 (2007)

24. Srinivasan, A., King, R.D., Muggleton, S., Sternberg, M.J.E.: Carcinogenesis predictions using ILP. In: Džeroski, S., Lavrač, N. (eds.) ILP 1997. LNCS, vol. 1297, pp. 273–287. Springer, Heidelberg (1997)
25. Srinivasan, A., Muggleton, S., Sternberg, M.J.E., King, R.D.: Theories for mutagenicity: A study in first-order and feature-based induction. Artif. Intell. **85**(1–2), 277–299 (1996)
26. Sun, Z., Fox, G.: Study on parallel svm based on mapreduce. In: Proceedings of the 18th International Conference on Parallel and Distributed Processing Techniques and Applications, pp. 16–19 (2012)
27. Zese, Riccardo, Bellodi, Elena, Lamma, Evelina, Riguzzi, Fabrizio, Aguiari, Fabiano: Semantics and inference for probabilistic description logics. In: Bobillo, Fernando, Carvalho, Rommel N., Costa, Paulo C.G., d'Amato, Claudia, Fanizzi, Nicola, Laskey, Kathryn B., Laskey, Kenneth J., Lukasiewicz, Thomas, Nickles, Matthias, Pool, Michael (eds.) URSW III. LNCS, vol. 8816, pp. 79–99. Springer, Heidelberg (2014)
28. Zese, R., Bellodi, E., Riguzzi, F., Lamma, E.: Tableau reasoners for probabilistic ontologies exploiting logic programming techniques. In: Bellodi, E., Bonfietti, A. (eds.) Proceedings of the Doctoral Consortium (DC) co-located with the 14th Conference of the Italian Association for Artificial Intelligence (AI*IA 2015). CEUR Workshop Proceedings, vol. 1485, pp. 1–6. Sun SITE Central Europe, Aachen (2015)

Meta-Interpretive Learning of Data Transformation Programs

Andrew Cropper[✉], Alireza Tamaddoni-Nezhad, and Stephen H. Muggleton

Department of Computing, Imperial College London, London, UK
a.cropper13@imperial.ac.uk

Abstract. Data transformation involves the manual construction of large numbers of special-purpose programs. Although typically small, such programs can be complex, involving problem decomposition, recursion, and recognition of context. Building such programs is common in commercial and academic data analytic projects and can be labour intensive and expensive, making it a suitable candidate for machine learning. In this paper, we use the meta-interpretive learning framework (MIL) to learn recursive data transformation programs from small numbers of examples. MIL is well suited to this task because it supports problem decomposition through predicate invention, learning recursive programs, learning from few examples, and learning from only positive examples. We apply Metagol, a MIL implementation, to both semi-structured and unstructured data. We conduct experiments on three real-world datasets: medical patient records, XML mondial records, and natural language taken from ecological papers. The experimental results suggest that high levels of predictive accuracy can be achieved in these tasks from small numbers of training examples, especially when learning with recursion.

1 Introduction

Suppose you are given a large number of patient records in a semi-structured format and are required to transform them to a given structured format. Figure 1 shows such a scenario relating to medical patient records, where (a) is the input and (b) is the desired output. To avoid manually transforming the records, you might decide to write a small program to perform this task. Figure 1c shows a Prolog program for this task which transforms the input to the output. However, manually writing this relatively simple program is somewhat laborious. In this paper, we show how such data transformation programs can be machine learned from a small number of input/output examples. Indeed, the program shown in Fig. 1c was learned from the given input/output examples by our system, described in Sect. 4. In this program, predicate invention is used to introduce the predicates $f1$ and $f2$ and the primitive background predicates *find_patient_id/2*, *find_int/2*, and *open_interval/4* are used to identify the various fields to be transposed from the input to the output.

In this paper, we investigate the problem of learning data transformation programs as an ILP application. In general, this is a challenging problem for

© Springer International Publishing Switzerland 2016
K. Inoue et al. (Eds.): ILP 2015, LNAI 9575, pp. 46–59, 2016.
DOI: 10.1007/978-3-319-40566-7_4

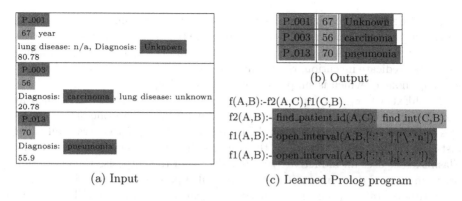

(a) Input

(b) Output

f(A,B):-f2(A,C),f1(C,B).

f2(A,B):- find_patient_id(A,C), find_int(C,B).

f1(A,B):- open_interval(A,B,[':',' '],['\','n']).

f1(A,B):- open_interval(A,B,[':',' '],[',',' ']).

(c) Learned Prolog program

Fig. 1. Transformation of medical records from semi-structured format

ILP which for example requires learning recursive rules from a small number of training examples. In order to address this problem, we use the recently developed meta-interpretive learning (MIL) framework [5,12,13] to learn data transformation programs. MIL differs from most state-of-the-art ILP approaches by supporting predicate invention for problem decomposition and the learning of recursive programs.

Using MIL to learn recursive programs has been demonstrated [4] to be particularly powerful in learning robot strategies, which are applicable to a potentially infinite set of initial/final state pairs, in contrast to learning non-recursive robot plans, applicable to only a specific initial/final state pair. We investigate learning recursive data transformation programs, applicable to a potentially infinite set of input/output examples.

The paper is arranged as follows. Section 2 describes related work. Section 3 describes the theoretical framework. Section 4 describes the transformation language used in the experiments. Section 5 describes experiments in learning recursive data transformation programs in three domains: medical patient records, XML mondial documents, and ecological scholarly papers. Finally, Sect. 6 concludes the paper and details future work.

2 Related Work

In [3] the authors compared statistical and relational methods to extract facts from a MEDLINE corpus. The primary limitation of the statistical approach, they state, is its inability to express the linguistic structure of the text; by contrast, the relational approach allows these features to be encoded as background knowledge, as parse trees, specifically. The relational approach used an ILP system similar to FOIL [14] to learn extraction rules. Similar works include [6], who used the ILP system Aleph [16] to learn rules to extract relations from a MEDLINE corpus; and [1], who used the ILP system FOIL to learn rules to extract relations from Nature and New Scientist articles. These works focused on constructing the appropriate problem representation, including determining the necessary linguistic features to be included in the background knowledge. In contrast to these approaches, we use the state-of-the-art ILP system Metagol, which supports predicate invention, the learning of recursive theories, and positive-only learning, none of which is supported by FOIL nor Aleph.

FlashExtract [8] is a framework to extract fields from documents, such as text files and web pages. In this framework, the user highlights one or two examples of each field in a document and FlashExtract attempts to extract all other instances of such fields, arranging them in a structured format, such as a table. FlashExtract uses an inductive synthesis algorithm to synthesise the extraction program using a domain-specific language built upon a pre-specified algebra of a few core operators (map, filter, merge, and pair). In contrast to FlashExtact, our approach allows for the inclusion of background knowledge.

Wu et al. [19] presented preliminary results on learning data transformation rules from examples. They demonstrated that specific string transformation rules can be learned from examples, given a grammar describing common user editing behaviors (i.e. insert, move, and delete). Their approach then uses a search algorithm to reduce the larger grammar space to a disjunction of subgrammar spaces (i.e. transformation rules) which are consistent with the examples. Depending on the grammar, the search could still generate many consistent transformations and they use a ranking algorithm to order transformation rules, e.g. based on the homogeneity of the transformed data. In their approach the set of transformation rules that the system can generate are pre-defined and not universal. By contrast, in our work, the transformation programs are not pre-defined and can be learned using predicate invention. Wu and Knoblock [18] recently extended their approach into a Programming-by-Example technique which iteratively learns data transformation programs by example. Their technique works by identifying previous incorrect subprograms and replacing them with correct subprograms. They demonstrated their technique on a set of string transformation problems and compared the results with the Flashfill approach [7] and Metagol$_{DF}$ [9]. While the overall aims of their approach are similar to ours, their approach does not support automatic problem decomposition using predicate invention, nor learning recursive programs. In this paper we also demonstrated our approach on a wider range of applications.

In [9], MIL was used to learn string transformation programs. In this approach, the authors perform transformations at the character level. However, this approach significantly increases the search space, and is unsuitable for learning from large input/output examples. By contrast, in this work, we look at the more general problem of learning data transformation programs, which are applicable to larger inputs/outputs and also a wider range of inputs/outputs. For instance, in Sect. 5, we apply our technique to relatively larger (55 kb) XML files.

ILP has been used in the past for the task of learning recursive rules from biological text. For example in [2], recursive patterns are discovered from biomedical text by inducing mutually dependent definitions of concepts using the ILP system ATRE. However, this is restricted in that they have a fixed number of slots in the output which need to be filled. By comparison, both the ecological and XML experiments in this paper show that we are not limited to this in our approach.

Similarly, ATRE has been used in [10] to learn a recursive logical theory of the ontology from a biological text corpus. However these ILP approaches cannot be easily extended for the general task of learning data transformation programs from examples. Both [2,10] have only been demonstrated on extraction of specific types of information from biological text, and did not attempt more general text transformation tasks of the kinds demonstrated in our paper. Moreover, these approaches were not shown to learn recursive rules from a small number of training examples and also do not support predicate invention.

3   Framework

MIL [12,13] is a form of ILP based on an adapted Prolog meta-interpreter. Whereas a standard Prolog meta-interpreter attempts to prove a goal by repeatedly fetching first-order clauses whose heads unify with a given goal, a MIL learner attempts to prove a set of goals by repeatedly fetching higher-order metarules (Fig. 3) whose heads unify with a given goal. The resulting meta-substitutions are saved in an abduction store, and can be re-used in later proofs. Following the proof of a set of goals, a hypothesis is formed by applying the meta-substitutions onto their corresponding metarules, allowing for a form of ILP which supports predicate invention and the learning of recursive theories.

General formal framework. In the general framework for data transformation we assume that the user provides examples E of how data should be transformed. Each example $e \in E$ consists of a pair $\langle d1, d2 \rangle$ where $d1 \in D1$ and $d2 \in D2$ are input and output data records respectively. Given background knowledge B, in the form of existing transformations and the user-provided examples E the aim of the learning is to generate a transformational function $\tau : D1 \rightarrow D2$ such that $B, \tau \models E$.

4   Implementation

Figure 2 shows the implementation of Metagol[1] as a generalised meta-interpreter [13], similar in form to a standard Prolog meta-interpreter.

Metagol works as follows. Metagol first tries to prove a goal deductively delegating the proof to Prolog (*call(Atom)*). Failing this, Metagol tries to unify the goal with the head of a metarule (*metarule(Name,Subs,(Atom :- Body))*) and to bind the existentially quantified variables in a metarule to symbols in the signature. Metagol saves the resulting *meta-substitutions* (*Subs*) in an abduction store and tries to prove the body goals of the metarule. After proving all goals, a Prolog program is formed by projecting the meta-substitutions onto their corresponding metarules.

---

[1] https://github.com/metagol/metagol.

```
prove([],H,H).
prove([Atom|Atoms],H1,H2):-
    prove_aux(Atom,H1,H3),
    prove(Atoms,H3,H2).
prove_aux(Atom,H,H):-
    call(Atom).
prove_aux(Atom,H1,H2):-
    member(sub(Name,Subs),H1),
    metarule(Name,Subs,(Atom :- Body)),
    prove(Body,H1,H2),
prove_aux(Atom,H1,H2):-
    metarule(Name,Subs,(Atom :- Body)),
    new_metasub(H1,sub(Name,Subs)),
    abduce(H1,H3,sub(Name,Subs)),
    prove(Body,H3,H2).
```

Fig. 2. Prolog code for Metagol, a generalised meta-interpreter

Name	Metarule	Order
Base	$P(x,y) \leftarrow Q(x,y)$	$P \succ Q$
Chain	$P(x,y) \leftarrow Q(x,z), R(z,y)$	$P \succ Q, P \succ R$
Curry	$P(x,y) \leftarrow Q(x,y,c_1,c_2)$	$P \succ Q$
TailRec	$P(x,y) \leftarrow Q(x,z), P(z,y)$	$P \succ Q, x \succ z \succ y$

Fig. 3. Metarules with associated ordering constraints, where $\succ$ is a pre-defined ordering over symbols in the signature. The letters P, Q, and R denote existentially quantified higher-order variables; x, y, and z denote universally quantified first-order variables; and c_1 and c_2 denote existentially quantified first-order variables

4.1 Transformation Language

In the transformation language, we represent the state as an *Input/Output* pair, where *Input* and *Output* are both character lists. The transformation language consists of two predicates *skip_to*/3 and *open_interval*/4. The *skip_to*/3 predicate is of the form *skip_to(A, B, Delim)*, where A and B are states, and *Delim* is a character list. This predicate takes a state $A = InputA/OutputA$ and skips to the delimiter *Delim* in *InputA* to form an output B = *InputB/OutputA*. For example, let $A = [i, n, d, u, c, t, i, o, n]/[]$ and *Delim* = $[u, c]$, then *skip_to(A, B, Delim)* is true where $B = [u, c, t, i, o, n]/[]$. The *open_interval*/4 predicate is of the form *open_interval(A, B, Start, End)*, where A and B are states, and *Start* and *End* are character lists. This predicate takes a state $A = InputA/OutputA$, finds a sublist in *InputA* denoted by *Start* and *End* delimiters, appends that sublist to *OutputA* to form *OutputB*, and skips all elements up to *End* delimiter in *InputA* to form *InputB*. For example, let $A = [i, n, d, u, c, t, i, o, n]/[]$, *Start* = $[n, d]$, and *End* = $[t, i]$, then open_interval(A,B,Start,End) is true where $B = [t, i, o, n]/[u, c]$.

5 Experiments

We now detail experiments[2] in learning data transformation programs. We test the following null hypotheses:

Null hypothesis 1. Metagol cannot learn data transformation programs with higher than default predictive accuracies.

Null hypothesis 2. $Metagol_R$ (with recursion) cannot learn data transformation programs with higher predictive accuracies than $Metagol_{NR}$ (without recursion).

To test these hypotheses, we apply our framework to three real-world datasets: XML mondial files, patient medical records, and ecological scholarly papers. To test null hypothesis 2, we learn using two versions of Metagol: $Metagol_R$ and $Metagol_{NR}$. Both versions use the *chain* and *curry* metarules, but $Metagol_R$ also uses the *tailrec* metarule.

We do not compare our results with other state-of-the-art ILP systems because they cannot support predicate invention, nor, crucially, the learning of recursive programs, and so such a comparison would be unfair.

5.1 XML Data Transformations

In this experiment, the aim is to learn programs to extract values from semi-structured data. We work with XML files, but the methods can be applied to other semi-structured mark-up languages, such as JSON.

Materials. The dataset[3] is a 1mb worldwide geographic database XML file which contains information about 231 countries, such as population, provinces, cities, etc. We split the file so that each country is a separate XML file. We consider the task of extracting all the city names for each country. Figure 4 shows three simplified positive examples, where the left column (input) is the XML file and the right column (output) is the city names to be extracted. Appendix A shows a full input example used in the experiments. Note the variations in the dataset, such as the varying number of cities (from 1 to over 30) and the differing XML structures. To generate training and testing examples, we wrote a Python script (included as Appendix B) to extract all the city names for each country. This was a non-trivial task for us, and it would be even more difficult for non-programmers, which supports the claim that this should be automated). The 231 pairings of a country XML file and the cities in that country form the positive examples. We do not, however, use all the positive examples as training examples because the amount of information varies greatly depending on the country. For instance, the file on the British crown dependency of Jersey is 406 bytes, whereas the file on the USA is 55 kb. We postulate that in a real-world setting a user is unlikely

---

[2] Experiments available at https://github.com/andrewcropper/ilp15-datacurate.
[3] http://www.cs.washington.edu/research/xmldatasets/www/repository.html#mondial.

Input	Output

`<country id='f0_136' name='Albania'`
`  capital='f0_1461'>`
` <name>Albania</name>`
` <city><name>Tirane</name></city>`
` <city><name>Shkoder</name></city>`
` <city><name>Durres</name></city>`
`</country>`
	Tirane, Shkoder, Durres

`<country id='f0_144' name='Andorra'`
`    capital='f0_1464'>`
` <name>Andorra</name>`
` <city><name>Andorra la Vella</name></city>`
`</country>`
	Andorra la Vella

`<country id='f0_149' name='Austria'`
`    capital='f0_1467'>`
` <name>Austria</name>`
` <province name="Burgenland">`
`  <city><name>Eisenstadt</name></city>`
` </province>`
` <province name="Vienna">`
`  <city><name>Vienna</name></city>`
` </province>`
`</country>`
	Eisenstadt, Vienna

Fig. 4. Three simplified XML transformation positive examples where all the city names have been extracted. Most of the XML has been removed for brevity, but is included in the experiments. The actual examples contain a lot more information and a full example is included as Appendix A

to manually annotate (i.e. label) a 55 kb file. Therefore, we only use country files less than 2 kb as positive training examples, of which there are 182. We do, however, use the larger country files for testing. To generate negative examples, for each XML file we extracted all k text entries and randomly selected j values to form the negative output, ensuring that the random sample did correspond to the expected output. Figure 5 displays an example negative instance.

Methods. For each m in the set $\{1, \ldots, 10\}$ we randomly select without replacement m positive and m negative training examples. The default predictive accuracy is therefore 50 %. We average predictive accuracies and learning times over 10 trials. We set a maximum solution length to 5.

Results. Figure 6 shows that Metagol learns solutions with higher than default predictive accuracies, refuting null hypothesis 1. Figure 6 also shows that learning with recursion results in higher predictive accuracies compared to learning without recursion, refuting null hypothesis 2. The difference in performance is because the non-recursive learner cannot handle the varying number of cities, whereas the recursive solution (Fig. 7) can handle any number of cities.

Input

`<country capital="f0_1557" name="Liechtenstein">`
`  <name>Liechtenstein</name>`
`  <city>`
`    <name>Vaduz</name>`
`    <population>27714</population>`
`  </city>`
`  <ethnicgroups>Italian</ethnicgroups>`
`  <ethnicgroups>Alemannic</ethnicgroups>`
`  <religions>Roman Catholic</religions>`
`  <religions>Protestant</religions>`
`</country>`

Output
`Italian, 27714, Vaduz, Liechtenstein, Alemannic, Protestant, Roman Catholic`

Fig. 5. Simplified XML transformation negative example

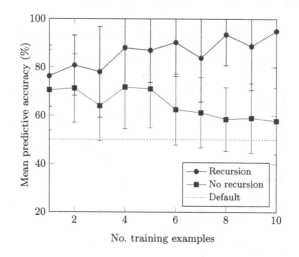

Fig. 6. XML learning performance when varying number of training examples

5.2 Ecological Scholarly Papers

In this experiment, the aim is to learn programs to extract relations from natural language taken from ecological scholarly papers.

Materials. The dataset contains 25 positive real-world examples of a natural language sentence paired with a list of values to be extracted. These examples are taken from ecological scholarly papers adopted from [17]. Figure 8 shows two example pairings where we want to extract the predator name, the predation term, and a list of all prey. We provide all species and predication terms in the background knowledge. No negative examples are provided, so we learn using positive examples only.

f(A,B):- f1(A,C), f(C,B).
f1(A,B):- open_interval(A,B,['a','m','e','>'],['<o','/','n','a']).
f(A,B):- skip_to(A,B,['<','/','n','a']).

Fig. 7. Example recursive solution learned by Metagol on the XML dataset

input 1:   This species also has a wide food range, but whereas Feronia melanaria took
         Coleoptera adults as the main item of the diet, Nebria brevicollis took spiders,
         Collembola, Coleoptera adults and larvae in equal number in the present study.
output 1: ['Nebria brevicollis', 'took', 'spiders', 'Collembola', 'Coleoptera adults and
         larvae']

input 2:   Bembidion lampros. This species is an important predator of cabbage root
         fly eggs (Hughes 1959; Coaker & Williams 1963) and it also feeds on Collembola,
         mites, pseudo-scorpions, earthworms and small bettles (Mitchell 1963a).
output 2: ['Bembidion lampros', 'predator', 'cabbage root fly eggs', 'Collembola',
         'mites', 'pseudo-scorpions', 'earthworms', 'small bettles']

Fig. 8. Two input/output examples from the ecological experiment

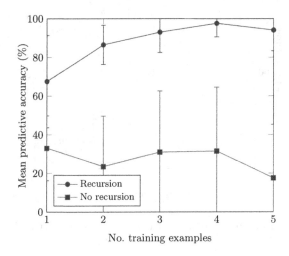

Fig. 9. Ecological learning performance when varying number of training examples

Methods. For each m in the set $\{1, \ldots, 10\}$, we randomly select without replacement m positive training examples and 10 positive testing examples. We average
predictive accuracies and learning times over 20 trials. We set a maximum solution length to 5.

Results. Figure 9 shows that learning with recursion significantly improves predicate accuracies compared to learning without recursion, again refuting null
hypothesis 2. Figure 10 shows an example learned recursive solution.

f(A,B):- f3(A,C), f2(C,B).
f3(A,B):- f2(A,C), find_predation(C,B).
f2(A,B):- find_species(A,B).
f2(A,B):- find_species(A,C), f2(C,B).

Fig. 10. Example recursive solution learned by Metagol on the ecological dataset

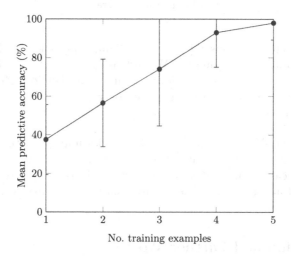

Fig. 11. Medical record learning performance when varying number of training examples

5.3 Patient Medical Records

In this experiment, the aim is to learn programs to extract values from patient medical records.

Materials. The dataset contains 16 positive patient medical records, modelled on real-world examples[4], paired with a list of values to be extracted. Figures 1a and 1b show simplified input/output example pairings. The experimental dataset, however, contains one additional input and output value, which is a floating integer value. We provide the following background predicates: find_int/2, find_float/2, and find_patient_id/2. We do not, however, provide a predicate to identify the diagnosis field, so Metagol must use the general purpose background predicates, described in Sect. 4, to learn a solution. To generate negative testing examples, we create a frequency distribution over input lengths from the training examples and a frequency distribution over words and punctuation in the training examples. To create a negative testing example input, we randomly select a length n from the length distribution and then randomly select with replacement n words or punctuation characters.

---

[4] http://www.ncbi.nlm.nih.gov/geo/query/acc.cgi?acc=GSE8581.

f(A,B):- f2(A,C), f2(C,B).
f2(A,B):- find_patient_id(A,C), find_int(C,B).
f2(A,B):- f1(A,C), find_float(C,B).
f1(A,B):- open_interval(A,B,[':',' '],['\','n']).
f1(A,B):- open_interval(A,B,[':',' '],[',',' ']).

Fig. 12. Learned program

Methods. For each m in the set $\{1, \ldots, 5\}$, we randomly select without replacement m positive training examples. We test using 20 positive and 20 negative testing examples. The default predictive accuracy is therefore 50 %. We average predictive accuracies and learning times over 20 trials. We set a maximum solution length to 5.

Results. Figure 11 shows that predictive accuracies improve with an increasing number of training examples, with over 80 % predictive accuracy from a single example. In all cases, the predictive accuracies of learned solutions are greater than the default accuracy, and thus null hypothesis 1 is refuted. Figure 12 shows an example solution.

6   Conclusion and Further Work

We have investigated learning programs which transform data from one format to another, and we have introduced a general framework for the problem. Our experiments on medical patient records, XML mondial files, and ecological natural language texts, indicate that MIL is capable of generating accurate recursive data transformation programs from small numbers of user-provided examples.

This paper provides an initial investigation into an important new group of ILP applications relevant to information extraction by example problems. Further work remains to achieve this potential.

Several issues need to be studied further in scaling-up the work reported. To begin with, although training data required will be small, owing to the requirements of user-provided annotation, test data will typically consist of large numbers of instances. Running Prolog hypotheses on such data will be time-consuming, and we would like to investigate generating hypotheses in a scripting language, such as Python.

For data transformation problems such as the ecological dataset we would also like to investigate the value of large-scale background knowledge, which might provide deeper natural language analysis based on dictionaries, tokenisation, part-of-speech tagging, and specialised ontologies.

We would also like to investigate Probabilistic ILP approaches [15], such as [11], which have the potential to not only provide probabilistic preferences over hypothesised programs, but also the potential of dealing with issues such as noise which are ubiquitous within real-world data. In the context of free-text data these approaches might also be integrated with finding highest probability parses.

Acknowledgements. The first author acknowledges the support of the BBSRC and Syngenta in funding his PhD Case studentship. The second author acknowledges the support from the IMI eTRIKS project. The third author would like to thank the Royal Academy of Engineering and Syngenta for funding his present 5 year Research Chair.

A   Appendix 1

```xml
<?xml version="1.0" encoding="UTF-8"?>
<country capital="f0_2148" car_code="GH" datacode="GH"
gdp_agri="47" gdp_ind="16" gdp_serv="37" gdp_total="25100"
government="constitutional democracy" id="f0_1269"
indep_date="06 03 1957" infant_mortality="80.3"
inflation="69" name="Ghana" population="17698272"
population_growth="2.29" total_area="238540">
  <name>Ghana</name>
  <city country="f0_1269" id="f0_2148" latitude="5.55" longitude="-0.2">
    <name>Accra</name>
    <population year="84">867459</population>
    <located_at type="sea" water="f0_38068" />
  </city>
  <city country="f0_1269" id="f0_16328">
    <name>Kumasi</name>
    <population year="84">376249</population>
  </city>
  <city country="f0_1269" id="f0_16333">
    <name>Cape Coast</name>
    <population year="84">57224</population>
  </city>
  <city country="f0_1269" id="f0_16338">
    <name>Tamale</name>
    <population year="84">135952</population>
  </city>
  <city country="f0_1269" id="f0_16343">
    <name>Tema</name>
    <population year="84">131528</population>
  </city>
  <city country="f0_1269" id="f0_16348">
    <name>Takoradi</name>
    <population year="84">61484</population>
  </city>
  <city country="f0_1269" id="f0_16353">
    <name>Sekondi</name>
    <population year="84">31916</population>
  </city>
  <ethnicgroups percentage="0.2">European</ethnicgroups>
  <ethnicgroups percentage="99.8">African</ethnicgroups>
  <religions percentage="30">Muslim</religions>
  <religions percentage="24">Christian</religions>
  <encompassed continent="f0_132" percentage="100" />
```

```
    <border country="f0_1189" length="548" />
    <border country="f0_1231" length="668" />
    <border country="f0_1422" length="877" />
</country>
```

B   Appendix 2

```
from xml.dom import minidom
doc = minidom.parse('mondial.xml')
countries = doc.getElementsByTagName('country')
i=1
for country in countries:
  cities = country.getElementsByTagName('city')
  names = [city.getElementsByTagName('name')[0].childNodes[0].data.strip()
           for city in cities]
  with open('parsed/output-{0}.txt'.format(i),'w') as f:
    f.write(','.join(names))
  i+=1
```

References

1. Aitken, J.S.: Learning information extraction rules: An inductive logic programming approach. In: ECAI, pp. 355–359 (2002)
2. Berardi, M., Malerba, D.: Learning recursive patterns for biomedical information extraction. In: Muggleton, S.H., Otero, R., Tamaddoni-Nezhad, A. (eds.) ILP 2006. LNCS (LNAI), vol. 4455, pp. 79–93. Springer, Heidelberg (2007)
3. Craven, M., Kumlien, J., et al.: Constructing biological knowledge bases by extracting information from text sources. ISMB **1999**, 77–86 (1999)
4. Cropper, A., Muggleton, S.H.: Learning efficient logical robot strategies involving composable objects. In: Proceedings of the 24th International Joint Conference Artificial Intelligence (IJCAI 2015), pp. 3423–3429. IJCAI (2015)
5. Cropper, A., Muggleton, S.H.: Logical minimisation of meta-rules within meta-interpretive learning. In: Davis, J., Ramon, J. (eds.) ILP 2014. LNCS, vol. 9046, pp. 62–75. Springer, Heidelberg (2015). doi:10.1007/978-3-319-23708-4_5
6. Goadrich, M., Oliphant, L., Shavlik, J.: Learning ensembles of first-order clauses for recall-precision curves: a case study in biomedical information extraction. In: Camacho, R., King, R., Srinivasan, A. (eds.) ILP 2004. LNCS (LNAI), vol. 3194, pp. 98–115. Springer, Heidelberg (2004)
7. Gulwani, S.: Automating string processing in spreadsheets using input-output examples. In: Proceedings of the 38th ACM SIGPLAN-SIGACT Symposium on Principles of Programming Languages, POPL 2011, Austin, TX, USA, 26–28 January 2011, pp. 317–330 (2011)
8. Le, V., Gulwani, S.: Flashextract: A framework for data extraction by examples. In: ACM SIGPLAN Notices, vol. 49, pp. 542–553. ACM (2014)
9. Lin, D., Dechter, E., Ellis, K., Tenenbaum, J.B., Muggleton, S.H.: Bias reformulation for one-shot function induction. In: Proceedings of the 23rd European Conference on Artificial Intelligence (ECAI 2014), pp. 525–530. IOS Press, Amsterdam (2014)

10. Manine, A.-P., Alphonse, E., Bessières, P.: Extraction of genic interactions with the recursive logical theory of an ontology. In: Gelbukh, A. (ed.) CICLing 2010. LNCS, vol. 6008, pp. 549–563. Springer, Heidelberg (2010)
11. Tamaddoni-Nezhad, A., Muggleton, S.: Stochastic refinement. In: Frasconi, P., Lisi, F.A. (eds.) ILP 2010. LNCS, vol. 6489, pp. 222–237. Springer, Heidelberg (2011)
12. Muggleton, S.H., Lin, D., Pahlavi, N., Tamaddoni-Nezhad, A.: Meta-interpretive learning: application to grammatical inference. Mach. Learn. **94**, 25–49 (2014)
13. Muggleton, S.H., Lin, D., Tamaddoni-Nezhad, A.: Meta-interpretive learning of higher-order dyadic datalog: predicate invention revisited. Mach. Learn. **100**(1), 49–73 (2015). doi:10.1007/s10994-014-5471-y
14. Quinlan, J.R., Cameron-Jones, R.M.: FOIL: a midterm report. In: Brazdil, P.B. (ed.) ECML 1993. LNCS, vol. 667. Springer, Heidelberg (1993)
15. De Raedt, L., Kersting, K.: Probabilistic inductive logic programming. In: De Raedt, L., Frasconi, P., Kersting, K., Muggleton, S.H. (eds.) Probabilistic Inductive Logic Programming. LNCS (LNAI), vol. 4911, pp. 1–27. Springer, Heidelberg (2008)
16. Srinivasan, A.: The Aleph Manual. University of Oxford, Oxford (2007)
17. Sunderland, K.D.: The diet of some predatory arthropods in cereal crops. J. Appl. Ecol. **12**(2), 507–515 (1975)
18. Bo, W., Knoblock, C.A.: An iterative approach to synthesize data transformation programs. In: Proceedings of the 24th International Joint Conference on Artificial Intelligence (IJCAI) (2015)
19. Bo, W., Szekely, P., Knoblock, C.A.: Learning data transformation rules through examples: preliminary results. In: Proceedings of the Ninth International Workshop on Information Integration on the Web, IIWeb 2012, pp. 8:1–8:6. ACM, New York, NY, USA (2012)

Statistical Relational Learning
with Soft Quantifiers

Golnoosh Farnadi[1,2]([✉]), Stephen H. Bach[3], Marjon Blondeel[4],
Marie-Francine Moens[2], Lise Getoor[5], and Martine De Cock[1,6]

[1] Department of Applied Mathematics, Computer Science and Statistics,
Ghent University, Ghent, Belgium
`golnoosh.farnadi@ugent.be`
[2] Department of Computer Science, Katholieke Universiteit Leuven, Leuven, Belgium
`sien.moens@cs.kuleuven.be`
[3] Statistical Relational Learning Group, University of Maryland, College Park, USA
`bach@cs.umd.edu`
[4] Ghent University Global Campus, Incheon, South Korea
`marjon.blondeel@ugent.be`
[5] Statistical Relational Learning Group, University of California, Santa Cruz, USA
`getoor@soe.ucsc.edu`
[6] Center for Data Science, University of Washington, Tacoma, USA
`mdecock@uw.edu`

Abstract. Quantification in statistical relational learning (SRL) is
either existential or universal, however humans might be more inclined
to express knowledge using soft quantifiers, such as "most" and "a few".
In this paper, we define the syntax and semantics of PSL^Q, a new SRL
framework that supports reasoning with soft quantifiers, and present
its most probable explanation (MPE) inference algorithm. To the best
of our knowledge, PSL^Q is the first SRL framework that combines soft
quantifiers with first-order logic rules for modeling uncertain relational
data. Our experimental results for link prediction in social trust networks
demonstrate that the use of soft quantifiers not only allows for a natural
and intuitive formulation of domain knowledge, but also improves the
accuracy of inferred results.

1 Introduction

Statistical relational learning (SRL) has become a popular paradigm for knowledge representation and inference in application domains with uncertain data
that is of a complex, relational nature. A variety of different SRL frameworks
has been developed over the last decade, based on ideas from probabilistic graphical models, first-order logic, and programming languages (see e.g., [11,21,26]).
Quantification in first-order logic is traditionally either existential ($\exists$) or universal ($\forall$). Given the strong roots of the existing SRL frameworks in (a subset of)
first-order logic as a knowledge representation language, it is no surprise that
these are the two kinds of quantifications that are known and commonly used in

© Springer International Publishing Switzerland 2016
K. Inoue et al. (Eds.): ILP 2015, LNAI 9575, pp. 60–75, 2016.
DOI: 10.1007/978-3-319-40566-7_5

SRL, even though in many application scenarios humans might be more inclined to express knowledge using softer quantifiers, such as *most* and *a few*.

For example, in models for social networks it is common to include the knowledge that the behaviour, beliefs, and preferences of friends all influence each other. How this information can be incorporated depends on the expressivity of the model. In a traditional probabilistic model, a dependency might be included for each pair of friends (corresponding to a universally quantified rule), each expressing the knowledge that it is more probable that two friends share a trait in common. An often cited example in SRL contexts describing smoking behaviour among friends is $\forall X \forall Y Friends(X, Y) \rightarrow (Smokes(X) \leftrightarrow Smokes(Y))$ [26]. This formula states that if two people are friends, then either both of them smoke or neither of them. In this case, the probability that a person smokes scales smoothly with the number of friends that smoke. However, many traits of interest might not behave this way, but instead exhibit "tipping points" in which having a trait only becomes more probable once *most* or *some* of one's friends have that trait (e.g., smoking behaviour). Expressing this dependency requires a soft quantifier, which none of the existing SRL frameworks allow.

What sets soft quantifiers apart from universal and existential quantification is that expressions that contain them are often true to a certain degree, as opposed to either being true or false. Indeed, the degree to which a statement such as "most of Bob's friends smoke" is true, increases with the percentage of smokers among Bob's friends. This increase is not necessarily linear; in fact, a common approach to compute the truth degree of soft quantified expressions is to map percentages to the scale $[0, 1]$ using non-decreasing piecewise linear functions [31]. Previous SRL work (e.g., [15,20,23]) has considered hard quantifiers with thresholds such as *at least k*. Soft quantifiers, on the other hand, do not impose such hard thresholds but allow smooth, gradual transitions from falsehood to truth.

Furthermore, the dependence of predicted probabilities on population size in relational models such as Markov logic networks (MLNs) and relational logistic regression is addressed in [16,22]. Soft quantifiers not only provide the flexibility of modeling complex relations, but their semantics also do not depend on the absolute population size. Hence soft quantifiers allow us to learn a model for some population size and apply the same model to another population size without the need for changes in the model, e.g. without introducing auxiliary variables to control whether the population size grows.

Many SRL applications could benefit from the availability of soft quantifiers. Collective document classification, for instance, relies on rules such as $\forall D \forall E \forall C (Cites(D, E) \wedge Class(D, C) \rightarrow Class(E, C))$ which expresses that if documents D and E are linked (e.g., by citation), and D belongs to class C, then E belongs to C [2]. Soft quantifiers would allow to classify a document based on *most* of its citing documents instead of one citing document. Similarly, in collaborative filtering, one can rely on the preferred products of a user to infer the behaviour of a similar user, i.e., $\forall U_1 \forall U_2 \forall J (Likes(U_1, J) \wedge Similar(U_1, U_2) \rightarrow Likes(U_2, J))$ [2]. Using soft quantifiers would allow to infer preferences of a user based on *most* of the behaviours of a similar user, or by comparing one user with *most* of the users similar to him.

In this paper we present the first SRL framework that combines soft quantifiers with first-order logic rules for modeling uncertain relational data. A brief overview of our framework is presented in [10]. We start from probabilistic soft logic (PSL) [1], an existing SRL framework that defines templates for hinge-loss Markov random fields [2], and extend it to a new framework which we call PSL^Q. As is common in SRL frameworks, in PSL a problem is defined by a set of logical rules using a finite set of atoms. However, unlike other SRL frameworks whose atoms are Boolean, atoms in PSL can take continuous values in the interval $[0,1]$. Intuitively, value 0 means *false* and value 1 means *true*, while any value $v \in [0,1]$ represents a partial degree of truth. PSL has been used in various domains with promising results, including trust propagation [13], drug-target interaction prediction [9], knowledge graph identification [25], semantic textual similarity computation [4] and sentiment analysis in a social network [29], among many others.

Our approach differs from existing research on quantifiers for logical reasoning in various ways. Studies on quantifiers in probabilistic logic settings deal with Boolean atoms [3,19,27], while in this paper atoms take on continuous values. The literature on fuzzy logic contains a fair amount of work on reasoning with continuous values (e.g., [6,24]), including the use of soft quantifiers [5], yet, to the best of our knowledge, there is no prior work on such soft quantifiers in SRL.

This paper makes three contributions. First, we introduce PSL^Q, a new SRL framework that supports reasoning with soft quantifiers, such as "most" and "a few". Second, because this expressivity pushes beyond the capabilities of PSL, we introduce new inference and weight learning algorithms for PSL^Q. Finally, as a proof of concept, we present a PSL^Q model that more accurately predicts trust in social networks than the current state-of-the-art approach.

2   PSL^Q: PSL with Soft Quantifiers

Definition 1. *An **atom** is an expression of the form $p(a_1, a_2, \ldots, a_n)$ where p is a **predicate symbol**, and each argument $a_1, a_2, \ldots, a_n$ is either a constant or a variable. The finite set of all possible substitutions of a variable to a constant for a particular variable a_i is called its **domain** D_{a_i}. If all variables in $p(a_1, a_2, \ldots, a_n)$ are substituted by some constant from their respective domain, then we call the resulting atom a **ground atom**. We call $\neg p(a_1, a_2, ..., a_n)$ a **negated atom** which is the negation of $p(a_1, a_2, ..., a_n)$.*

Definition 2. *A **quantifier expression** is an expression of the form*

$$Q(V, F_1[V], F_2[V]) \tag{1}$$

*where Q is a **soft quantifier**, and $F_1[V]$ and $F_2[V]$ are formulas containing the variable V. A formula is an atom or a negation, conjunction or disjunction of formulas. A **grounded quantifier expression** is obtained by instantiating all variables with constants from their domains except for V.*

Consider as an example the two formulas $Knows(X,T)$ and $Trusts(X,T)$, then $Most(T, Knows(X,T), Trusts(X,T))$ is a quantifier expression. By substituting X with Alice, we obtain the grounded quantifier expression:

$Most(T, Knows(Alice, T), Trusts(Alice, T))$, which can be read as "Alice trusts most of the people she knows".

Definition 3. *A **PSL^Q** **model** consists of a collection of PSL^Q rules. A **PSL^Q** **rule** r is an expression of the form:*

$$\lambda_r : T_1 \wedge T_2 \wedge \ldots \wedge T_k \rightarrow H_1 \vee H_2 \vee \ldots \vee H_l \tag{2}$$

where $T_1, T_2, \ldots, T_k, H_1, H_2, \ldots, H_l$ are atoms, negated atoms, quantifier expressions or negated quantifier expressions and $\lambda_r \in \mathbb{R}^+ \cup \{\infty\}$ is the weight of the rule r. We call $T_1 \wedge T_2 \wedge \ldots \wedge T_k$ the body of r (r_{body}), and $H_1 \vee H_2 \vee \ldots \vee H_l$ the head of r (r_{head}). Grounding a PSL^Q rule means instantiating all the variables with constants from their domain except for the variables V in quantifier expressions $Q(V, F_1[V], F_2[V])$.

Remark 1. *A PSL model, i.e., a set of PSL rules, is a PSL^Q model without quantifier expressions. The first 9 rules in Table 1 are an example of a PSL^Q model without quantifier expressions, or a PSL model, while rules $10 - 14$ in Table 1 are examples of PSL^Q rules with quantifier expressions. In the remainder of this paper, we use the term 'standard PSL' to refer to PSL without quantifiers as defined in [1].*

An **interpretation** I is a mapping that associates a truth value $I(s) \in [0, 1]$ to each ground atom s. For example, $I(Knows(Alice, Bob)) = 0.7$ indicates that Alice knows Bob to degree 0.7. The interpretation of PSL^Q rules is based on Łukasiewicz logic [17]. Conjunction $\wedge$ is interpreted by the Łukasiewicz t-norm ($\tilde{\wedge}$), disjunction $\vee$ by the Łukasiewicz t-conorm ($\tilde{\vee}$), and negation $\neg$ by the Łukasiewicz negator ($\tilde{\neg}$), which are defined as follows. For $m, n \in [0, 1]$ we have: $m \tilde{\wedge} n = \max(0, m + n - 1)$, $m \tilde{\vee} n = \min(m + n, 1)$ and $\tilde{\neg} m = 1 - m$. The $\tilde{\ }$ indicates the relaxation over Boolean values. We can extend the interpretation of atoms to more complex formulas in Łukasiewicz logic as follows. Given an interpretation I, and ϕ_1 and ϕ_2 formulas, we have $I(\phi_1 \wedge \phi_2) = I(\phi_1) \tilde{\wedge} I(\phi_2)$, $I(\phi_1 \vee \phi_2) = I(\phi_1) \tilde{\vee} I(\phi_2)$ and $I(\neg \phi_1) = \tilde{\neg} I(\phi_1)$.

The interpretation of quantifier expressions in PSL^Q relies on quantifier mappings.

Definition 4. *A **quantifier mapping** $\tilde{Q}$ is a $[0, 1] \rightarrow [0, 1]$ mapping. If $\tilde{Q}$ is non-decreasing and satisfies the boundary conditions $\tilde{Q}(0) = 0$ and $\tilde{Q}(1) = 1$, it is called a **coherent** quantifier mapping [8].*

We assume that for every soft quantifier Q an appropriate quantifier mapping $\tilde{Q}$ can be defined, i.e. a function that represents the meaning of Q. Using two thresholds $\alpha \in [0, 1]$ and $\beta \in [0, 1]$, where $\alpha < \beta$, the following equation defines a parametrized family of such quantifier mappings:

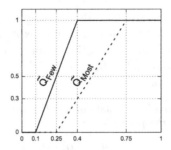

Fig. 1. Examples of quantifier mappings

$$\tilde{Q}_{[\alpha,\beta]}(x) = \begin{cases} 0 & \text{if } x < \alpha \\ \frac{x-\alpha}{\beta-\alpha} & \text{if } \alpha \le x < \beta \\ 1 & \text{if } x \ge \beta \end{cases} \tag{3}$$

Figure 1 depicts a possible coherent quantifier mapping for the soft quantifier "a few" as $\tilde{Q}_{Few} = \tilde{Q}_{[0.1,0.4]}$ and for the soft quantifier "most" as $\tilde{Q}_{Most} = \tilde{Q}_{[0.25,0.75]}$. Note how $\tilde{Q}_{Few}$ is more relaxed than $\tilde{Q}_{Most}$. For example, using these mappings, the statement "a few friends of Bob smoke" is true to degree 1 as soon as 40 % of Bob's friends are smokers, while 75 % of Bob's friends are required to be smokers for the statement "most friends of Bob smoke" to be fully true. The evaluation section contains a detailed analysis on the effect of the choice of the thresholds α and β on the results obtained with MPE inference. In practice friendship is not necessarily a black-and-white matter, i.e., people can be friends to varying degrees. For instance, $I(Friend(Bob, Alice)) = 1$ and $I(Friend(Bob, Chris)) = 0.2$ denote that under interpretation I, Alice is a very close friend of Bob, while Chris is a more distant friend. Similarly, Chris might be a heavy smoker, while Alice might be only a light smoker. All these degrees can and should be taken into account when computing the truth degree of statements such as "a few friends of Bob smoke" and "most friends of Bob smoke".

We define the interpretation of a grounded quantifier expression based on the Zadeh approach [31]. Zadeh suggested to calculate the truth value of "Q A's are Bs", with $A : \mathcal{U} \to [0,1]$ and $B : \mathcal{U} \to [0,1]$ fuzzy sets in a universe $\mathcal{U}$, as:

$$\tilde{Q}\left(\frac{|A \cap B|}{|A|}\right)$$

where $A \cap B$ is a fuzzy set defined as: $A \cap B : \mathcal{U} \to [0,1] : u \mapsto A(u) \tilde{\wedge} B(u)$. In this expression, the **cardinality** of a fuzzy set $S : \mathcal{U} \to [0,1]$ is defined as $|S| = \sum_{u \in \mathcal{U}} S(u)$.

Definition 5. *For a given interpretation I, **the interpretation of a grounded quantifier expression** $Q(V, F_1[V], F_2[V])$ is defined as*

$$I(Q(V, F_1[V], F_2[V])) = \tilde{Q}\left(\frac{\sum_{x \in D_V} I(F_1(x)) \tilde{\wedge} I(F_2(x))}{\sum_{x \in D_V} I(F_1(x))}\right) \tag{4}$$

with $\tilde{Q}$ a quantifier mapping modeling Q and D_V the domain of V.

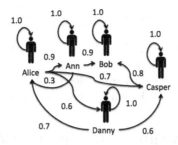

Fig. 2. Sample trust network between five users

Example 1. *Let's consider an interpretation I in a sample trust network as shown in Fig. 2. Nodes represent users and each edge represents the trust relation between two users. Since a trust relation is asymmetric, the direction of the trust relation is shown with an arrow. The degree of the trust links are shown with a value under/above the links, e.g., $I(Trusts(Alice, Ann)) = 0.9$.*

To calculate $I(Most(X, Trusts(Alice, X), Trusts(X, Bob)))$, we calculate:
$\sum_x(I(Trusts(Alice, x))\tilde{\wedge} I(Trusts(x, Bob))) = 1.3$ *and* $\sum_x I(Trusts(Alice, x)) = 3.2$, *thus we have* $\tilde{Q}\left(\frac{1.3}{3.2}\right) \sim \tilde{Q}(0.41)$. *By using the quantifier expression mapping of "most" in Fig. 1, we obtain* $\tilde{Q}_{[0.25, 0.75]}(0.41) = 0.32$. *Thus, under interpretation I, "most trustees of Alice trust Bob" to degree 0.32.*

Remark 2. *In Łukasiewicz logic, the formula $\phi_1 \to \phi_2$ where $\to$ is implication, is logically equivalent to the formula $\neg\phi_1 \vee \phi_2$, thus the interpretation of a grounded PSL^Q rule r is as follows:*

$$I(r) = I(r_{body} \to r_{head}) = \tilde{\neg}I(r_{body})\tilde{\vee}I(r_{head}) \tag{5}$$

Definition 6. *The distance to satisfaction $d_r(I)$ of a rule r under an interpretation I is defined as:*

$$d_r(I) = \max\{0, I(r_{body}) - I(r_{head})\} \tag{6}$$

By using Remark 2, one can show that a rule r is fully satisfied, i.e. satisfied to degree 1, when the truth value of its head is at least as high as the truth value of its body. Thus, the closer the interpretation of a grounded rule is to 1, the smaller its distance to satisfaction.

A PSL^Q model, i.e., a set of PSL^Q rules, induces a distribution over interpretations I. Let R be the set of all grounded rules, then the probability density function is:

$$f(I) = \frac{1}{Z}\exp[-\sum_{r \in R}\lambda_r(d_r(I))^p] \tag{7}$$

where λ_r is the weight of rule r, Z is a normalization constant

$$Z = \int_I \exp[-\sum_{r \in R}\lambda_r(d_r(I))^p]$$

and $p \in \{1, 2\}$. These probabilistic models are instances of hinge-loss Markov random fields (HL-MRF). For further explanation we refer to [2]. Choosing $p = 1$ (i.e., linear) favors interpretations that completely satisfy one rule at the expense of higher distance from satisfaction for conflicting rules, and $p = 2$ favors interpretations that satisfy all rules to some degree (i.e., quadratic). Note that in Sect. 3 we only consider $p = 1$, since by [2] the results can be extended for $p = 2$.

3   Inference and Weight Learning in PSL^Q

Expressing soft quantifiers pushes beyond the capabilities of inference and weight learning methods in PSL. In this section, we introduce new methods for inference based on the most probable explanation inference method (MPE inference) and weight learning with maximum-likelihood estimation (MLE) in PSL^Q.

3.1   Inference

The goal of MPE "most probable explanation" inference is to find the most probable truth assignments I_{MPE} of unknown ground atoms given the evidence which is defined by the interpretation I. Let X be all the evidence, i.e., X is the set of ground atoms such that $\forall x \in X, I(x)$ is known, and let Y be the set of ground atoms such that $\forall y \in Y, I(y)$ is unknown. Then we have

$$I_{MPE}(Y) = arg\ max_{I(Y)} P(I(Y)|I(X)) \tag{8}$$

and by Eq. 7 it follows that the goal of optimization is to minimize the weighted sum of the distances to satisfaction of all rules.

Remark 3. *Suppose we want to optimize a $f : [0,1]^n \to [0,1]$ function consisting of applications of only piecewise linear functions, fractions of piecewise linear functions, min $: [0,1]^2 \to [0,1]$ and max $: [0,1]^2 \to [0,1]$. We can transform such an optimization problem as follows. For every expression of the form $\min(\phi, \psi)$, we introduce a variable $v_{\min(\phi,\psi)}$ and add the constraints $0 \le \phi, \psi, v_{\min(\phi,\psi)} \le 1$, $v_{\min(\phi,\psi)} \le \phi$ and $v_{\min(\phi,\psi)} \le \psi$. Similarly, for every expression of the form $\max(\phi, \psi)$, we introduce a variable $v_{\max(\phi,\psi)}$ and add the constraints $0 \le \phi, \psi, v_{\max(\phi,\psi)} \le 1$, $v_{\max(\phi,\psi)} \ge \phi$ and $v_{\max(\phi,\psi)} \ge \psi$. Define the function g as the original function f but all minima and maxima are replaced by their corresponding variables. Optimizing f is then equivalent to optimizing g under these constraints.*

By the particular piecewise linear form of $d_r(I)$ (see Eq. 6) and Remark 3, standard PSL's underlying HL-MRFs have log concave density functions and hence finding an MPE assignment is a convex optimization problem, which is solvable in polynomial time. Standard PSL only supports linear constraints to preserve convexity. Hence, standard PSL potentially can support linear aggregates.

Definition 7. *An **aggregate** is a $[0,1]^n \rightarrow [0,1]$ mapping. If it is a linear mapping, it is called a **linear aggregate**, otherwise it is called a **non-linear aggregate**.*

As an example, $f : [0,1]^n \rightarrow [0,1] : (t_1, .., t_n) \mapsto \frac{t_1 + t_2 + ... + t_n}{n}$ is a linear aggregate. A PSL^Q program allows expressions that contain quantifier expressions. Since the interpretation of a grounded quantifier expression (see Eq. 4) is based on a non-linear aggregate, finding a MPE assignment of a PSL^Q program with quantifier expressions is beyond the capabilities of the standard PSL MPE-solver. To deal with this, we will first categorize different types of grounded quantifier expressions, given the interpretation I denoting the evidence.

Definition 8. *A grounded quantifier expression $Q(V, F_1[V], F_2[V])$, where for every $s \in D_V$, it holds that all ground atoms in the formulas $F_1[s]$ and $F_2[s]$ are in X, is called a **fully observed grounded quantifier expression (FOQE)**.*

For instance, in a social network where the age and the friends of all users are known, by grounding $Most(X, Friend(A, X), Young(X))$, we obtain FOQEs. Note that for a FOQE $Q(V, F_1[V], F_2[V])$, we have that $I(Q(V, F_1[V], F_2[V]))$ is a known value in $[0, 1]$.

Definition 9. *A grounded quantifier expression $Q(V, F_1[V], F_2[V])$, where for every $s \in D_V$, it holds that all ground atoms in the formula $F_1[s]$ are in X and there exists $t \in D_V$ such that at least one ground atom in the formula $F_2[t]$ is in Y, is called a **partially observed grounded quantifier expression of type one (POQE$^{(1)}$)**.*

Suppose all friendship relations are known and the goal is to infer the age of all users based on the age of some, then by grounding $Most(X, Friend(A, X), Young(X))$, we obtain POQE$^{(1)}$s. Note that for a POQE$^{(1)}$ $Q(V, F_1[V], F_2[V])$, we have that $I(Q(V, F_1[V], F_2[V])) = \tilde{Q}(f(Y))$ where f is a piecewise linear function in variables belonging to Y.

Definition 10. *A grounded quantifier expression $Q(V, F_1[V], F_2[V])$, for which there exists $t \in D_V$ such that at least one ground atom in the formula $F_1[t]$ is in Y, is called a **partially observed grounded quantifier expression of type two (POQE$^{(2)}$)**.*

Note that for a POQE$^{(2)}$ $Q(V, F_1[V], F_2[V])$, we have that $I(Q(V, F_1[V], F_2[V])) = \tilde{Q}(f(Y))$ where f is a fraction of piecewise linear functions in variables belonging to Y. In link prediction applications, such as trust link prediction, we mostly deal with POQE$^{(2)}$s. By grounding the rules $10 - 14$ in Table 1 using unknown trust relations, we obtain complex examples of POQE$^{(2)}$s.

In the following proposition we give an equivalent definition for the membership function in Eq. 3. By applying Remark 3 we will then be able to show that a PSL^Q program can be transformed to a linear fractional program (LFP).

Algorithm 1. Iterative MPE inference in PSL^Q

Require: PSL^Q program P, evidence variables X and random variables Y

1: $R \leftarrow \emptyset$
2: $I^{(0)}(Y) \leftarrow 0$
3: **for** $i := 1$ **to** k **do**
4:     **for** $r \in P$ **do**
5:         $R_g \leftarrow$ **ground**(r)
6:         **for** $r_g \in R_g$ **do**
7:             **for** every Q of type POQE$^{(2)}$ in r_g **do**
8:                 $I(Q) \leftarrow \tilde{Q}(I(X) \cup I^{i-1}(Y))$
9:             **end for**
10:             $d_{r_g}(I) \leftarrow 1 - I(r_g)$
11:             **if not** $d_{r_g}(I) = 0$ **then**
12:                 $R \leftarrow R \cup r_g$
13:             **end if**
14:         **end for**
15:     **end for**
16:     $f(I) \leftarrow$ **generate**(R)
17:     $G(I) \leftarrow$ **transform**$(f(I))$
18:     $I^{(i)}(Y) \leftarrow$ **optimize**$(G(I))$
19: **end for**

Proposition 1. *The membership-function defined in Eq. 3 where $\alpha \in [0,1]$, $\beta \in [0,1]$, and $\alpha < \beta$ can be rewritten as:*

$$\tilde{Q}_{[\alpha,\beta]}(x) = \max(0, \frac{x-\alpha}{\beta-\alpha}) + \min(\frac{x-\alpha}{\beta-\alpha}, 1) - \frac{x-\alpha}{\beta-\alpha} \qquad (9)$$

After grounding a PSL^Q program we can obtain a mixture of FOQEs, POQE$^{(1)}$s and POQE$^{(2)}$s. Recall that for a FOQE $Q(V, F_1[V], F_2[V])$, we have that $I(Q(V, F_1[V], F_2[V])) \in [0,1]$. On the other hand, for a POQE$^{(1)}$ $Q(V, F_1[V], F_2[V])$, we have that $I(Q(V, F_1[V], F_2[V])) = \tilde{Q}(f(Y))$ where f is a piecewise linear function in variables belonging to Y and for a POQE$^{(2)}$ $Q(V, F_1[V], F_2[V])$, we have that $I(Q(V, F_1[V], F_2[V])) = \tilde{Q}(g(Y))$ where g is a fraction of piecewise linear functions in variables belonging to Y. By applying Proposition 1 and Remark 3 it then follows that a PSL^Q program using piecewise linear quantifier mappings such as in Eq. 3 can be transformed to a linear fractional program (LFP). Note that this is only a worst case scenario: if the grounded PSL^Q program has no POQE$^{(2)}$'s then we obtain a linear program. We can then use a transformation similar to the approach of Isbell and Marlow [14] to replace a LFP by a set of linear programs by establishing a convergent iterative process. The linear program at each iteration is determined by optimization of the linear program at the previous iteration.

The algorithm we propose for the MPE inference (Algorithm 1) starts by initializing all random variables to zero (i.e., line 2). Then, an iterative process starts by grounding all rules in a PSL^Q program (i.e., line 3–5). For every grounded quantifier expression Q of type POQE$^{(2)}$, the value of Q is initialized by calculating the value over the known values ($I(X)$) and the current setting of the unknown values ($I^{(0)}(Y)$). In the algorithm, we use the notation $\tilde{Q}(I(X) \cup I^{i-1}(Y))$ to denote this new interpretation of Q at iteration i (i.e., line 7–9). For each rule r_g we then calculate the distance to satisfaction (i.e., line 10). Note that $I(r_g)$ and hence also $d_{r_g}(I)$ can be piecewise linear functions in Y, but here $d_{r_g}(I)$ does not contain fractions of piecewise linear functions since we calculate values for the POQE$^{(2)}$s. Next, we exclude the satisfied grounded rules (i.e., we exclude rules r_g such that $d_{r_g}(I) = 0$) from the optimization since their values will not change the optimization task (i.e., line 11–13). For the optimization task, $f(I)$ (Eq. 7) is calculated using the distance to satisfaction of all grounded rules (i.e., line 16). Since $f(I)$ does not contain fractions of piecewise linear functions, it can be transformed to a linear program (i.e., line 17). Finally, the inner optimization in PSL^Q is solved with PSL's scalable, parallelizable message-passing inference algorithm [2] (i.e., line 18). In each iteration, the values of Qs get updated by the most probable assignment of random variables in the previous iteration ($I(X) \cup I^{(i-1)}(Y)$) (i.e., line 8). This process is iteratively repeated for a fixed number of times (i.e., k).

3.2 Weight Learning

The goal of weight learning based on maximum likelihood estimation (MLE) is to maximize the log likelihood of the rules' weight based on the training data in Eq. 7. Hence, the partial derivatives of log likelihood with respect to λ_i of rule $r_i \in R$ are

$$-\frac{\delta \log(f(I))}{\delta \lambda_i} = E_\lambda [\sum_{r \in R_g i} (d_r(I))^p] - \sum_{r \in R_g i} (d_r(I))^p \tag{10}$$

with E_λ the expected value under the distribution defined by λ, and $R_g i$ is the set of grounded rules of rule r_i. The optimization is based on the voted perception algorithm [7], in which approximation is done by taking fixed-length steps in the direction of gradient and averaging the points after all steps; out of the scope steps are projected back into the feasible region. To make the approximation tractable, a MPE approximation is used that replaces the expectation in the gradient with the corresponding values in the MPE state. We use our proposed MPE approach for transforming POQE$^{(1)}$s and POQE$^{(2)}$s in our MLE algorithm. We omit the pseudocode of the MLE algorithm for a PSL^Q program to save space.

4 Evaluation: Trust Link Prediction

Studies have shown that people tend to rely more on recommendations from people they trust than on online recommender systems which generate recommendations based on anonymous people similar to them. This observation has generated a rising interest in trust-enhanced recommendation systems [28]. The recommendations generated by these systems are based on an (online) trust network, in which members of the community express whether they trust or distrust each other. In practice these networks are sparse because most people are connected to relatively few others. Trust-enhanced recommendation systems therefore rely on link prediction.

In [13], trust relations between social media users are modeled and predicted using a PSL model based on the structural balance theory [12]. Structural balance theory implies the transitivity of a relation between users. Based on this theory, users are more prone to trust their neighbors in the network rather than unknown other users. In [2][1], Bach et al. evaluated the PSL model based on the structural balance theory on data from *Epinions*[2], an online consumer review site in which users can indicate whether they trust or distrust each other. Throughout this section, we will use the same sample of Epinions [18]. The sample dataset includes 2,000 users with 8,675 relations, namely 7,974 trust relations and only 701 distrust relations.

We systematically perform 8-fold cross-validation and to evaluate the results, we use three metrics, *AUC*: the area under the receiver operating characteristic curve, *PR+*: the area under the precision-recall curves for trust relations, and *PR-*: the area under the precision-recall curves for distrust relations. In each fold, we first learn the weights of the rules based on 7/8 of the trust network and then apply the learned model on the remaining 1/8 to infer the Bach et al. used the model of [13] which is composed of twenty PSL rules in order to predict the degree of trust between two individuals. Sixteen rules from these rules encode possible stable triangular structures involving the two individuals and a third one. For example, an individual is likely to trust people his or her friends trust. The model of [13] is used to predict unobserved truth-values of $Trusts(A, B)$ for pairs of individuals. The results of this model are shown in the first line in Table 2.

In this paper, we propose a model based on 4 transitive rules (rules 1–4 in Table 1) and one rule which models the cyclic relation between 3 users (rule 5 in Table 1). Rules 6–9 in Table 1 are complementary rules for which we refer to [13] for further explanation. The atom $Average(\{Trusts\})$ in rules 8 and 9 is a constant which refers to the global average value of observed trust scores. This atom is useful for the disconnected parts of the trust network without any known trust relation. These four rules are also used in the PSL model of [2]. To investigate whether we can improve the accuracy of the predictions by introducing rules with soft quantifier expressions, we construct PSL^Q rules based on a triad relation over a set of users instead of a third party (rules 10–14). The full PSL^Q model then consists of all rules displayed in Table 1.

We examine what happens when changing the thresholds for the quantifier mappings $\tilde{Q}$ (Eq. 3). We have investigated ten different quantifier mappings by changing the values of α and β by steps of 0.25. In this way, we obtain ten different PSL^Q programs. For every program, we applied Algorithm 1 for all $k \in \{1, 2, \ldots, 10\}$. Note that for $k = 1$ the output of the MPE inference is equivalent to the output generated by a PSL^Q model with only FOQEs by ignoring the

---

[1] Source code available at http://psl.umiacs.umd.edu.
[2] www.epinions.com.

Table 1. PSL^Q model for trust link prediction

	Transitive rules
(R#1)	$Knows(A,B) \land Trusts(A,B) \land Knows(B,C) \land Trusts(B,C) \land Knows(A,C) \rightarrow Trusts(A,C)$
(R#2)	$Knows(A,B) \land \neg Trusts(A,B) \land Knows(B,C) \land Trusts(B,C) \land Knows(A,C) \rightarrow \neg Trusts(A,C)$
(R#3)	$Knows(A,B) \land Trusts(A,B) \land Knows(B,C) \land \neg Trusts(B,C) \land Knows(A,C) \rightarrow \neg Trusts(A,C)$
(R#4)	$Knows(A,B) \land \neg Trusts(A,B) \land Knows(B,C) \land \neg Trusts(B,C) \land Knows(A,C) \rightarrow Trusts(A,C)$
	Cyclic rule
(R#5)	$Knows(A,B) \land Trusts(A,B) \land Knows(B,C) \land Trusts(B,C) \land Knows(C,A) \rightarrow Trusts(C,A)$
	Complementary rules
(R#6)	$Knows(A,B) \land Knows(B,A) \land Trusts(B,A) \rightarrow Trusts(A,B)$
(R#7)	$Knows(A,B) \land Knows(B,A) \land \neg Trusts(B,A) \rightarrow \neg Trusts(A,B)$
(R#8)	$Knows(A,B) \land Average(\{Trusts\}) \rightarrow Trusts(A,B)$
(R#9)	$Knows(A,B) \land Trusts(A,B) \rightarrow Average(\{Trusts\})$
	PSL^Q rules based on the transitive rules
(R#10)	$Q(X, Knows(A,X) \land Trusts(A,X), Knows(X,C) \land Trusts(X,C)) \land Knows(A,C) \rightarrow Trusts(A,C)$
(R#11)	$Q(X, Knows(A,X) \land \neg Trusts(A,X), Knows(X,C) \land C30 \land Knows(A,C) \rightarrow \neg Trusts(A,C)$
(R#12)	$Q(X, Knows(A,X) \land Trusts(A,X), Knows(X,C) \land \neg Trusts(X,C)) \land Knows(A,C) \rightarrow \neg Trusts(A,C)$
(R#13)	$Q(X, Knows(A,X) \land \neg Trusts(A,X), Knows(X,C) \land \neg Trusts(X,C)) \land Knows(A,C) \rightarrow Trusts(A,C)$
	PSL^Q rule based on the cyclic rule
(R#14)	$Q(X, Knows(A,X) \land Trusts(A,X), Knows(X,C) \land Trusts(X,C)) \land Knows(C,A) \rightarrow Trusts(C,A)$

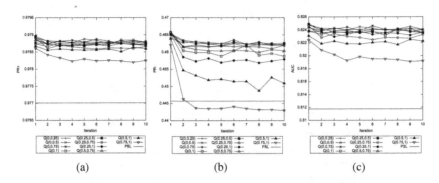

(a)   (b)   (c)

Fig. 3. (a) PR+, (b) PR−, and (c) AUC of changing α and β in the quantifier mapping $\tilde{Q}$

unknown values. Figure 3 presents changes of the three metrics of these ten PSL^Q models with different quantifier mappings. All ten PSL^Q models outperform the PSL model (shown with a line) in all iterations and in all three metrics, except for the PSL^Q model with $\tilde{Q}_{[0.75,1]}$ in $PR-$ after the first two iterations. An explanation for this is the fact that people trust/distrust a third party as soon as a few/some of their trusted/distrusted friends trust/distrust that person and not most of them, i.e., more than 75 %. Interestingly, by decreasing both α and β values, results get better. The model with the best predicting scores is PSL^Q with $\tilde{Q}_{[0,0.25]}$ as a quantifier mapping representing "a few" (see Table 2).

Figure 4 emphasizes the importance of the PSL^Q rules with quantifier expressions (rules 10–14) after the weight learning phase. Bars represent average and error bars represent minimum and maximum weights of the rules learned in 8 folds for the PSL^Q model with quantifier mapping $\tilde{Q}_{[0,0.25]}$. These results show that using soft quantifiers not only improves the accuracy of trust and distrust predictions but also that the rules containing soft quantifiers, i.e. rules 10–14, play a major part in this by dominating all other rules in terms of weight. In these experiments, we used one quantifier mapping for all the quantifiers in a PSL^Q program; however it is possible to use different mapping functions for each quantifier expression in a PSL^Q model, which is an interesting direction for future research.

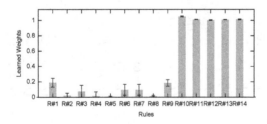

Fig. 4. Learned values of the weight of the 14 rules of the PSL^Q model

Table 2. Values with a ∗ are statistically significant with a rejection threshold of 0.05 and values in bold are statistically significant with a rejection threshold of 0.1 using a paired t-test w.r.t. the PSL model [2]. Distrust prediction is more challenging than trust prediction (i.e., PR- values are overall lower than PR+ values) because of the unbalanced nature of the data (7,974 trust vs. 701 distrust relations)

Method	PR+	PR-	AUC
PSL	0.977	0.446	0.812
PSL^Q $(\tilde{Q}_{[0,0.25]})$, $(k=1)$	0.979*	0.467*	0.825*
PSL^Q $(\tilde{Q}_{[0,0.25]})$, $(k=10)$	0.979*	0.463	0.824*

5 Conclusion

In this paper, we have introduced PSL^Q, the first SRL framework that supports reasoning with soft quantifiers, such as "most" and "a few". PSL^Q is a powerful and expressive language to model uncertain relational data in an intuitive way. Since this expressivity pushed beyond the capabilities of existing PSL-MPE solvers, we have introduced and implemented new inference and weight learning algorithms that can handle rules with soft quantifiers. We have shown how the higher expressivity of PSL^Q can lead to better results in practice by extending an existing PSL model for link prediction in social trust networks with rules that contain soft quantifiers. We have presented the effects of using different interpretations of soft quantifiers in our trust model. As a next step, we want to include an automatic way of learning the best interpretation for each quantifier expression in a PSL^Q model. Besides trust link prediction, many other applications could benefit from the use of soft quantifiers. Exploring the effects of using soft quantifiers in PSL^Q models for other AI applications is therefore another promising research direction. Furthermore, in addition to the approach of Zadeh that we have used in this paper, other approaches for soft quantifiers have been proposed, most notably Yager's OWA-operators [30]; we plan to investigate them in our future work.

Acknowledgements. We would like to thank the anonymous reviewers for their helpful comments and suggestions. This work was funded in part by the SBO-program of the Flemish Agency for Innovation by Science and Technology (IWT-SBO-Nr. 110067) and NSF grant IIS1218488. Any opinions, findings, and conclusions or recommendations expressed in this material are those of the author(s) and do not necessarily reflect the views of the NSF.

References

1. Bach, S.H., Broecheler, M., Huang, B., Getoor, L.: Hinge-loss markov random fields and probabilistic soft logic. [cs.LG] (2015). arXiv:1505.04406
2. Bach, S.H., Huang, B., London, B., Getoor, L.: Hinge-loss Markov random fields: convex inference for structured prediction. In: Proceedings of the Uncertainty in Artificial Intelligence (UAI) (2013)

3. Beltagy, I., Erk, K.: On the proper treatment of quantifiers in probabilistic logic semantics. In: Proceedings of the 11th International Conference on Computational Semantics (IWCS), p. 140 (2015)
4. Beltagy, I., Erk, K., Mooney, R.J.: Probabilistic soft logic for semantic textual similarity. In: Proceedings of the 52nd Annual Meeting of the Association for Computational Linguistics (ACL), pp. 1210–1219 (2014)
5. Bobillo, F., Straccia, U.: fuzzyDL: an expressive fuzzy description logic reasoner. In: Proceedings of the IEEE International Conference on Fuzzy Systems (FUZZ-IEEE), pp. 923–930 (2008)
6. Cao, T.H., Rossiter, J.M., Martin, T.P., Baldwin, J.F.: On the implementation of fril++ for object-oriented logic programming with uncertainty and fuzziness. In: Bouchon-Meunier, B., Gutiérrez-Ríos, J., Magdalena, L., Yager, R.R. (eds.) Technologies for Constructing Intelligent Systems 2. Studies in Fuzziness and Soft Computing, vol. 90, pp. 393–406. Springer, Heidelberg (2002)
7. Collins, M.: Discriminative training methods for hidden Markov models: theory and experiments with perceptron algorithms. In: Proceedings of the International Conference on Empirical methods in Natural Language Processing (ACL), pp. 1–8 (2002)
8. Delgado, M., Sánchez, D., Vila, M.A.: Fuzzy cardinality based evaluation of quantified sentences. Int. J. Approximate Reason. **1**, 23–66 (2000)
9. Fakhraei, S., Huang, B., Raschid, L., Getoor, L.: Network-based drug-target interaction prediction with probabilistic soft logic. IEEE/ACM Trans. Comput. Biol. Bioinform. **11**(5), 775–787 (2014)
10. Farnadi, G., Bach, S.H., Moens, M.F., Getoor, L., De Cock, M.: Extending psl with fuzzy quantifiers. In: Proceedings of the Fourth International Workshop on Statistical Relational, AI at AAAI (StarAI) (2014)
11. Getoor, L., Taskar, B.: Introduction to Statistical Relational Learning. MIT press, Cambridge (2007)
12. Heider, F.: The Psychology of Interpersonal Relations. Wiley, New York (1958)
13. Huang, B., Kimmig, A., Getoor, L., Golbeck, J.: A flexible framework for probabilistic models of social trust. In: Greenberg, A.M., Kennedy, W.G., Bos, N.D. (eds.) SBP 2013. LNCS, vol. 7812, pp. 265–273. Springer, Heidelberg (2013)
14. Isbell, J.R., Marlow, W.H.: Attrition games. Naval Res. Logistics Q. **3**(1–2), 71–94 (1956)
15. Jain, D., Barthels, A., Beetz, M.: Adaptive Markov logic networks: learning statistical relational models with dynamic parameters. In: Proceedings of the European Conference on Artificial Intelligence (ECAI), pp. 937–942 (2010)
16. Kazemi, S.M., Buchman, D., Kersting, K., Natarajan, S., Poole, D.: Relational logistic regression. In: Proceedings of the International Conference on Principles of Knowledge Representation and Reasoning (KR) (2014)
17. Klir, G., Yuan, B.: Fuzzy Sets and Fuzzy Logic. Prentice Hall, New Jersey (1995)
18. Leskovec, J., Huttenlocher, D., Kleinberg, J.: Signed networks in social media. In: Proceedings of the 28th ACM Conference on Human Factors in Computing Systems (CHI) (2010)
19. Lowd, D., Domingos, P.: Recursive random fields. In: Proceedings of the International Joint Conference on Artificial Intelligence (IJCAI), pp. 950–955 (2007)
20. Milch, B., Zettlemoyer, L.S., Kersting, K., Haimes, M., Kaelbling, L.P.: Lifted probabilistic inference with counting formulas. In: Proceedings of the International Conference on Artificial Intelligence (AAAi) **8**, 1062–1068 (2008)
21. Muggleton, S., De Raedt, L.: Inductive Logic Programming: Theory and Methods. J. Logic Program. **19**(20), 629–679 (1994)

22. Poole, D., Buchman, D., Kazemi, S.M., Kersting, K., Natarajan, S.: Population size extrapolation in relational probabilistic modelling. In: Straccia, U., Calì, A. (eds.) SUM 2014. LNCS, vol. 8720, pp. 292–305. Springer, Heidelberg (2014)
23. Poole, D., Buchman, D., Natarajan, S., Kersting, K.: Aggregation and population growth: the relational logistic regression and Markov logic cases. In: Proceedings of the International Workshop on Statistical Relational AI at UAI (StarAI) (2012)
24. Prade, H., Richard, G., Serrurier, M.: Learning first order fuzzy logic rules. In: De Baets, B., Kaynak, O., Bilgiç, T. (eds.) IFSA 2003. LNCS, vol. 2715. Springer, Heidelberg (2003)
25. Pujara, J., Miao, H., Getoor, L., Cohen, W.: Knowledge graph identification. In: Alani, H., Kagal, L., Fokoue, A., Groth, P., Biemann, C., Parreira, J.X., Aroyo, L., Noy, N., Welty, C., Janowicz, K. (eds.) ISWC 2013, Part I. LNCS, vol. 8218, pp. 542–557. Springer, Heidelberg (2013)
26. Richardson, M., Domingos, P.: Markov logic networks. Mach. Learn. **62**(1–2), 107–136 (2006)
27. Van den Broeck, G., Meert, W., Darwiche, A.: Skolemization for weighted first-order model counting (2013). arXiv:1312.5378
28. Victor, P., Cornelis, C., De Cock, M.: Trust and recommendations. In: Ricci, F., Rokach, L., Shapira, B., Kantor, P.B. (eds.) Recommender Systems Handbook, pp. 645–675. Springer, Heidelberg (2011)
29. West, R., Paskov, H.S., Leskovec, J., Potts, C.: Exploiting social network structure for person-to-person sentiment analysis. Trans. Assoc. Comput. Linguist. (TACL) **2**, 297–310 (2014)
30. Yager, R.R.: On ordered weighted averaging aggregation operators in multicriteria decision making. IEEE Trans. Syst. Man Cybern. (IEEE SMC) **1**, 183–190 (1988)
31. Zadeh, L.A.: A computational approach to fuzzy quantifiers in natural languages. Comput. Math. Appl. **1**, 149–184 (1983)

Ontology Learning from Interpretations in Lightweight Description Logics

Szymon Klarman[1,2(✉)] and Katarina Britz[2,3]

[1] Department of Computer Science, Brunel University London, Uxbridge, UK
szymon.klarman@gmail.com
[2] CSIR Centre for Artificial Intelligence Research, Pretoria, South Africa
[3] Department of Information Science, Stellenbosch University,
Stellenbosch, South Africa

Abstract. Data-driven elicitation of ontologies from structured data is a well-recognized knowledge acquisition bottleneck. The development of efficient techniques for (semi-)automating this task is therefore practically vital — yet, hindered by the lack of robust theoretical foundations. In this paper, we study the problem of learning Description Logic TBoxes from interpretations, which naturally translates to the task of ontology learning from data. In the presented framework, the learner is provided with a set of positive interpretations (i.e., logical models) of the TBox adopted by the teacher. The goal is to correctly identify the TBox given this input. We characterize the key constraints on the models that warrant finite learnability of TBoxes expressed in selected fragments of the Description Logic $\mathcal{EL}$ and define corresponding learning algorithms.

1 Introduction

In the advent of the Web of Data and various "e-" initiatives, such as e-science, e-health, e-governance, etc., the focus of the classical knowledge acquisition bottleneck becomes ever more concentrated around the problem of constructing rich and accurate ontologies enabling efficient management of the existing abundance of data [1]. Whereas the traditional understanding of this bottleneck has been associated with the necessity of developing ontologies *ex ante*, in a top-down, data-agnostic manner, this seems to be currently evolving into a new position, recently dubbed the knowledge reengineering bottleneck [2]. In this view, the contemporary challenge is to, conversely, enable data-driven approaches to ontology design — methods that can make use and make sense of the existing data, be it readily available on the web or crowdsourced, leading to elicitation of the ontological commitments implicitly present on the data-level. Even though the development of such techniques and tools, which could help (semi-)automate thus characterized ontology learning processes, becomes vital in practice, the robust theoretical foundations for the problem are still rather limited. This work

This work was funded in part by the National Research Foundation under Grant no. 85482.

K. Inoue et al. (Eds.): ILP 2015, LNAI 9575, pp. 76–90, 2016.
DOI: 10.1007/978-3-319-40566-7_6

is an attempt at establishing exactly such foundations and focuses on some key theoretical issues towards this goal.

We study the problem of learning *Description Logic* (DL) TBoxes from interpretations, which naturally translates to the task of ontology learning from data. DLs are a popular family of knowledge representation formalisms [3], which have risen to prominence as, among others, the logics underpinning different profiles of the Web Ontology Language OWL[1]. In this paper, we focus on the lightweight DL $\mathcal{EL}$ [4] and some of its more specific fragments. This choice is motivated, on the one hand, by the interesting applications of $\mathcal{EL}$, especially as the logic behind OWL 2 $\mathcal{EL}$ profile, while on the other, by its relative complexity, which enables us to make interesting observations from the learning perspective. Our learning model is a variant of learning from positive interpretations (i.e., from models of the target theory) — a generally established framework in the field of inductive logic programming [5,6]. In our scenario, the goal of the learner is to correctly identify the target TBox $\mathcal{T}$ given a finite set of its finite models. Our overarching interest lies in algorithms warranting effective learnability in such setting with no or minimum supervision. Our key research questions and contributions are therefore concerned with the identification of specific languages and conditions on the learning input under which such algorithms can be in principle defined.

In the following two sections, we introduce DL preliminaries and discuss the adopted learning model. In Sect. 4, we identify two interesting fragments of $\mathcal{EL}$, called $\mathcal{EL}^{\mathrm{rhs}}$ and $\mathcal{EL}^{\mathrm{lhs}}$, which satisfy some basic necessary conditions enabling finite learnability, and at the same time, we show that full $\mathcal{EL}$ does not meet that same requirement. In Sect. 5, we devise a generic algorithm which correctly identifies $\mathcal{EL}^{\mathrm{rhs}}$ and $\mathcal{EL}^{\mathrm{lhs}}$ TBoxes from finite data, employing a basic equivalence oracle. Further, in case of $\mathcal{EL}^{\mathrm{rhs}}$, we significantly strengthen this result by defining an algorithm which makes no such calls to an oracle, and thus supports fully unsupervised learning. In Sect. 6, we compare our work to related contributions, in particular to the framework of learning TBoxes from entailment queries, by Konev et al. [7,8]. We conclude in Sect. 7 with an overview of interesting open problems.

2   Description Logic Preliminaries

The language of the Description Logic (DL) $\mathcal{EL}$ [4] is given by (1) a vocabulary $\Sigma = (N_C, N_R)$, where N_C is a set of concept names (i.e., unary predicates, e.g., Father, Woman) and N_R a set of role names (i.e., binary predicates, e.g., hasChild, likes), and (2) the following set of constructors for defining complex concepts, which shall be divided into two groups:

$$\mathcal{EL}:\quad C, D ::= \top \mid A \mid C \sqcap D \mid \exists r.C$$
$$\mathcal{L}^{\sqcap}:\quad C, D ::= \top \mid A \mid C \sqcap D$$

where $A \in N_C$ and $r \in N_R$. Concept $\top$ denotes all individuals in the domain, $C \sqcap D$ the class of individuals that are instances of both C and D, and $\exists r.C$

---

[1] See http://www.w3.org/TR/owl2-profiles/.

describes all individuals that are related to some instance of C via the role r. The set of $\mathcal{L}^\sqcap$ concepts naturally captures the propositional part of $\mathcal{EL}$. The *depth* of a subconcept D in C is the number of existential restrictions within the scope of which D remains. The *depth of a concept* C is the depth of its subconcept with the greatest depth in C. Every $\mathcal{L}^\sqcap$ concept is trivially of depth 0.

A *concept inclusion* (or a *TBox axiom*) is an expression of the form $C \sqsubseteq D$, stating that all individuals of type C are D. We sometimes write $C \equiv D$ as an abbreviation for two inclusions: $C \sqsubseteq D$ and $D \sqsubseteq C$. For instance, axioms (i) and (ii) below state, respectively, that (i) the class of mothers consists of all and only those individuals who are women and have at least one child, (ii) while every individual of type Father_of_boy is a father and has at least one male child:

$$\text{Mother} \equiv \text{Woman} \sqcap \exists \text{hasChild}.\top \qquad (i)$$
$$\text{Father\_of\_boy} \sqsubseteq \text{Father} \sqcap \exists \text{hasChild}.\text{Man} \qquad (ii)$$

A TBox (or *ontology*) is a finite set of such concept inclusions in a particular language fragment. The language fragments considered in this paper are classified according to the type of restrictions imposed on the syntax of concepts C and D in the concept inclusions $C \sqsubseteq D$ permitted in the TBoxes:

$\mathcal{EL}$: C and D are both $\mathcal{EL}$ concepts;
$\mathcal{EL}^{\text{rhs}}$: C is an $\mathcal{L}^\sqcap$ concept and D an $\mathcal{EL}$ concept;
$\mathcal{EL}^{\text{lhs}}$: C is an $\mathcal{EL}$ concept and D an $\mathcal{L}^\sqcap$ concept;
$\mathcal{L}^\sqcap$: C and D are both $\mathcal{L}^\sqcap$ concepts

For instance, a TBox consisting of axioms (i) and (ii) above, belongs to language $\mathcal{EL}$, as it in fact contains some $\mathcal{EL}^{\text{rhs}}$ axioms (Mother $\sqsubseteq$ Woman $\sqcap$ $\exists$hasChild.$\top$ and (ii)) as well as one $\mathcal{EL}^{\text{lhs}}$ axiom (Woman $\sqcap$ $\exists$hasChild.$\top$ $\sqsubseteq$ Mother).

The semantics of DL languages is defined through interpretations of the form $\mathcal{I} = (\Delta^\mathcal{I}, \cdot^\mathcal{I})$, where $\Delta^\mathcal{I}$ is a non-empty *domain of individuals* and $\cdot^\mathcal{I}$ is an *interpretation function* mapping each $A \in N_C$ to a subset $A^\mathcal{I} \subseteq \Delta^\mathcal{I}$ and each $r \in N_R$ to a binary relation $r^\mathcal{I} \subseteq \Delta^\mathcal{I} \times \Delta^\mathcal{I}$. The interpretation function is inductively extended over complex expressions according to the fixed semantics of the constructors:

$$\top^\mathcal{I} = \Delta^\mathcal{I}$$
$$(C \sqcap D)^\mathcal{I} = \{x \in \Delta^\mathcal{I} \mid x \in C^\mathcal{I} \cap D^\mathcal{I}\}$$
$$(\exists r.C)^\mathcal{I} = \{x \in \Delta^\mathcal{I} \mid \exists y : (x,y) \in r^\mathcal{I} \wedge y \in C^\mathcal{I}\}$$

An interpretation $\mathcal{I}$ *satisfies* a concept inclusion $C \sqsubseteq D$ ($\mathcal{I} \models C \sqsubseteq D$) *iff* $C^\mathcal{I} \subseteq D^\mathcal{I}$. Whenever $\mathcal{I}$ satisfies all axioms in a TBox $\mathcal{T}$ ($\mathcal{I} \models \mathcal{T}$), we say that $\mathcal{I}$ is a *model* of $\mathcal{T}$. Interpretations and models defined in this way are in fact usual Kripke structures, which can be naturally represented as labelled graphs, with nodes representing individuals in the domain, edges — roles, and labels — the interpretations of concept and role names, respectively. For instance, the three graphs in Fig. 1 all represent possible models of the TBox consisting of axioms (i) and (ii) above:

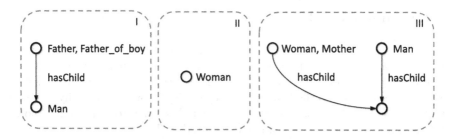

Fig. 1. Sample DL models.

Note that, as there are no a priori restrictions imposed on the number of domain individuals, a DL TBox might in general have infinitely many models of possibly infinite size. For a set of interpretations $\mathcal{S}$, we write $\mathcal{S} \models C \sqsubseteq D$ to denote that every interpretation in $\mathcal{S}$ satisfies $C \sqsubseteq D$. We say that $\mathcal{T}$ *entails* $C \sqsubseteq D$ $(\mathcal{T} \models C \sqsubseteq D)$ *iff* every model of $\mathcal{T}$ satisfies $C \sqsubseteq D$. Two TBoxes $\mathcal{T}$ and $\mathcal{H}$ are (logically) *equivalent* $(\mathcal{T} \equiv \mathcal{H})$ *iff* they have the same sets of models.

A *pointed interpretation* $(\mathcal{I}, d)$ is a pair consisting of a DL interpretation $\mathcal{I} = (\Delta^{\mathcal{I}}, \cdot^{\mathcal{I}})$ and an individual $d \in \Delta^{\mathcal{I}}$, such that every $e \in \Delta^{\mathcal{I}}$ different from d is reachable from d through some role composition in $\mathcal{I}$. By a slight abuse of notation, given an arbitrary DL interpretation $\mathcal{I}$ and an individual $d \in \Delta^{\mathcal{I}}$, we write $(\mathcal{I}, d)$ to denote the largest subset $\mathcal{I}'$ of $\mathcal{I}$ such that $(\mathcal{I}', d)$ is a pointed interpretation. If it is clear from the context, we refer to pointed interpretations and pointed models simply as interpretations and models. We say that $(\mathcal{I}, d)$ is a model of a concept C *iff* $d \in C^{\mathcal{I}}$; it is a model of C w.r.t. $\mathcal{T}$ whenever also $\mathcal{I} \models \mathcal{T}$.

An interpretation $(\mathcal{I}, d)$ can be *homomorphically embedded* in an interpretation $(\mathcal{J}, e)$, denoted as $(\mathcal{I}, d) \mapsto (\mathcal{J}, e)$, *iff* there exists a mapping $h : \Delta^{\mathcal{I}} \mapsto \Delta^{\mathcal{J}}$, satisfying the following conditions:

- $h(d) = e$,
- if $(a, b) \in r^{\mathcal{I}}$ then $(h(a), h(b)) \in r^{\mathcal{J}}$, for every $a, b \in \Delta^{\mathcal{I}}$ and $r \in N_R$,
- if $a \in A^{\mathcal{I}}$ then $h(a) \in A^{\mathcal{J}}$, for every $a \in \Delta^{\mathcal{I}}$ and $A \in N_C$.

A model $(\mathcal{I}, d)$ of C (w.r.t. $\mathcal{T}$) is called *minimal iff* it can be homomorphically embedded in every other model of C (w.r.t. $\mathcal{T}$). It is well-known that $\mathcal{EL}$ concepts and TBoxes always have such minimal models (unique up to homomorphic embeddings) [9]. As in most modal logics, arbitrary $\mathcal{EL}$ models can be unravelled into equivalent tree-shaped models. Finally, we observe that due to a tight relationship between the syntax and semantics of $\mathcal{EL}$, every tree-shaped interpretation $(\mathcal{I}, d)$ can be viewed as an $\mathcal{EL}$ concept $C_{\mathcal{I}}$, such that $(\mathcal{I}, d)$ is a minimal model of $C_{\mathcal{I}}$. Formally, we set $C_{\mathcal{I}} = C(d)$, where for every $e \in \Delta^{\mathcal{I}}$ we let $C(e) = \top \sqcap A(e) \sqcap \exists(e)$, with $A(e) = \bigsqcap\{A \in N_C \mid e \in A^{\mathcal{I}}\}$ and $\exists(e) = \bigsqcap_{(r,f) \in N_R \times \Delta^{\mathcal{I}} \text{ s.t. } (e,f) \in r^{\mathcal{I}}} \exists r.C(f)$. In that case we call $C_{\mathcal{I}}$ the *covering concept* for $(\mathcal{I}, d)$. For instance, the covering concept for model I in Fig. 1 is $\top \sqcap$ Father $\sqcap$ Father_of_boy $\sqcap \exists$hasChild.$(\top \sqcap$ Man$)$, which can be simplified as Father $\sqcap$ Father_of_boy $\sqcap \exists$hasChild.(Man).

3   Learning Model

The learning model studied in this paper is a variant of learning from positive interpretations [5,6]. In our setting, the teacher fixes a *target TBox* T, whose set of all models is denoted by $\mathcal{M}(T)$. Further, the teacher presents a set of examples from $\mathcal{M}(T)$ to the learner, whose goal is to correctly identify T based on this input. The learning process is conducted relative to a mutually known DL language $\mathcal{L}$ and a finite vocabulary Σ_T used in T.

In principle, $\mathcal{M}(T)$ contains sufficient information in order to enable correct identification of T, as the following correspondence implies:

$$\mathcal{M}(T) \models C \sqsubseteq D \; \textit{iff} \; T \models C \sqsubseteq D, \; \text{for every } C \sqsubseteq D \text{ in } \mathcal{L}.$$

However, as $\mathcal{M}(T)$ might consist of infinitely many models of possibly infinite size, the teacher cannot effectively present them all to the learner. Instead, the teacher must confine him- or herself to certain finitely presentable subset of $\mathcal{M}(T)$, called the *learning set*. For the sake of clarity, we focus here on the simplest case when learning sets consist of finitely many finite models.[2] Formally, we summarize the learning model with the following definitions.

Definition 1 (TIP). *A TBox Identification Problem (TIP) is a pair $(T, \mathcal{S})$, where T is a TBox in a DL language $\mathcal{L}$ and $\mathcal{S}$, called the* learning set*, is a finite set of finite models of T.*

Definition 2 (Learner, identification). *For a DL language $\mathcal{L}$, a learner is a computable function G, which for every set $\mathcal{S}$ over Σ_T returns a TBox in $\mathcal{L}$ over Σ_T. Learner G correctly identifies T on $\mathcal{S}$ whenever $G(\mathcal{S}) \equiv T$.*

Definition 3 (Learnability). *For a DL language $\mathcal{L}$, the class of TBoxes expressible in $\mathcal{L}$ is* learnable *iff there exists a learner G such that for every TBox T in $\mathcal{L}$ there exists a learning set $\mathcal{S}$ on which G correctly identifies T. It is said to be* finitely learnable *whenever it is learnable from finite learning sets only.*

We are primarily interested here in the notion of finite learnability, as it provides a natural formal foundation for the task of ontology learning from data. By data, in the DL context, we understand collections of atomic concept and role assertions over domain individuals (e.g., Father(john), hasChild(john, mary)), which under certain assumptions regarding their structuring with respect to the background ontology can be seen as models of that ontology and, consequently, as potentially valuable learning sets. Figure 2 presents an example of a TIP with a finite learning set, which consists of a single model of the assumed ontology. The key question is then what formal criteria must this set satisfy to warrant correct identification of the ontology constraining it. To this end we employ the

---

[2] An alternative, more general approach can be defined in terms of specific fragments of models. Such generalization, which lies beyond the scope of this paper, is essential when the learning problem concerns languages without finite model property.

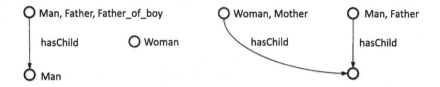

Mother ≡ Woman ⊓ ∃hasChild.⊤
Father ≡ Man ⊓ ∃hasChild.⊤
Father_of_boy ≡ Father ⊓ ∃hasChild.Man

Fig. 2. A sample TIP with an $\mathcal{EL}$ TBox and a finite learning set

basic *admissibility condition*, characteristic also of other learning frameworks [10], which ensures that the learning set is sufficiently rich to enable precise discrimination between the correct hypothesis and all the incorrect ones.

Definition 4 (Admissibility). *A TIP $(\mathcal{T}, \mathcal{S})$ is* admissible *iff for every $C \sqsubseteq D$ in $\mathcal{L}$ such that $\mathcal{T} \not\models C \sqsubseteq D$ there exists $\mathcal{I} \in \mathcal{S}$ such that $\mathcal{I} \not\models C \sqsubseteq D$.*

For the target TBox $\mathcal{T}$, let $\mathcal{T}^{\not\models}$ be the set of all concept inclusions in $\mathcal{L}$ that are not entailed by $\mathcal{T}$, i.e., $\mathcal{T}^{\not\models} = \{C \sqsubseteq D \text{ in } \mathcal{L} \mid \mathcal{T} \not\models C \sqsubseteq D\}$. The admissibility condition requires that for every $C \sqsubseteq D \in \mathcal{T}^{\not\models}$, the learning set $\mathcal{S}$ must contain a "counterexample" for it, i.e., an individual $d \in \Delta^{\mathcal{I}}$, for some $\mathcal{I} \in \mathcal{S}$, such that $d \in C^{\mathcal{I}}$ and $d \notin D^{\mathcal{I}}$. Consequently, any learning set must contain such counterexamples to all elements of $\mathcal{T}^{\not\models}$, or else, the learner might never be justified to exclude some of these concept inclusions from the hypothesis. If it was possible to represent them finitely we could expect that ultimately the learner can observe all of them and correctly identify the TBox. In the next section, we investigate this prospect formally in different fragments of $\mathcal{EL}$.

4 Finite Learning Sets

As argued in the previous section, to enable finite learnability of $\mathcal{T}$ in a given language $\mathcal{L}$, the relevant counterexamples to all the concept inclusions not entailed by $\mathcal{T}$ must be presentable within a finite learning set $\mathcal{S}$. Firstly, we can immediately observe that this requirement is trivially satisfied for $\mathcal{L}^{\sqcap}$. Clearly, $\mathcal{L}^{\sqcap}$ can only induce finitely many different concept inclusions (up to logical equivalence) on finite vocabularies, such as $\Sigma_{\mathcal{T}}$. Hence, the set $\mathcal{T}^{\not\models}$ can always be finitely represented (up to logical equivalence) and it is straightforward to finitely present counterexamples to all its members. For more expressive fragments of $\mathcal{EL}$, however, this cannot be assumed in general, as the $\exists r.C$ constructor induces infinitely many concepts. One negative result comes with the case of $\mathcal{EL}$ itself, as demonstrated in the next theorem.

Theorem 1 (Finite learning sets in $\mathcal{EL}$). *Let $\mathcal{T}$ be a TBox in $\mathcal{EL}$. There exists no finite set $\mathcal{S}$ such that $(\mathcal{T}, \mathcal{S})$ is admissible.*

The full proof of this and subsequent results is included in an online technical report [11]. The argument rests on the following lemma. Let $(\mathcal{T}, \mathcal{S})$ be an admissible TIP and C a concept. By $\mathcal{S}(C)$ we denote the set of all models $(\mathcal{I}, d)$ of C w.r.t. $\mathcal{T}$ such that $\mathcal{I} \in \mathcal{S}$. By $\bigcap \mathcal{S}(C)$ we denote the intersection of all these models, i.e., the model $(\mathcal{J}, d)$, such that:

1. $(\mathcal{J}, d) \mapsto (\mathcal{I}, d)$ for every $(\mathcal{I}, d) \in \mathcal{S}(C)$,
2. for every other model $(\mathcal{J}', d)$ such that $(\mathcal{J}', d) \mapsto (\mathcal{I}, d)$ for every $(\mathcal{I}, d) \in \mathcal{S}(C)$ and $(\mathcal{J}, d) \mapsto (\mathcal{J}', d)$, it is the case that $(\mathcal{J}', d) \mapsto (\mathcal{J}, d)$.

Lemma 1 (Minimal model lemma). *Let $(\mathcal{T}, \mathcal{S})$ be an admissible TIP for $\mathcal{T}$ in $\mathcal{EL}$ (resp. in $\mathcal{EL}^{rhs}$), and C be an $\mathcal{EL}$ (resp. $\mathcal{L}^{\sqcap}$) concept. Whenever $\mathcal{S}(C)$ is non-empty then $\bigcap \mathcal{S}(C)$ is a minimal model of C w.r.t. $\mathcal{T}$.*

Given the lemma, we consider a concept inclusion of type:

$$\tau_n := \underbrace{\exists r. \ldots \exists r.}_{n} \top \sqsubseteq \underbrace{\exists r. \ldots \exists r. \exists r.}_{n+1} \top$$

Suppose $\tau_n \in \mathcal{T}^{\not\models}$ for some $n \in \mathbb{N}$. Since by the admissibility condition a counterexample to τ_n must be present in $\mathcal{S}$, it must be the case that $\mathcal{S}(C) \neq \emptyset$, where C is the left-hand-side concept in τ_n. By the lemma and the definition of a minimal model, it is easy to see that $\mathcal{S}$ must contain a finite chain of individuals of length exactly $n + 1$, as depicted below:

Finally, since there can always exist some $n \in \mathbb{N}$, such that $\tau_m \in \mathcal{T}^{\not\models}$ for every $m \geq n$, we see that the joint size of all necessary counterexamples in such cases must inevitably be also infinite. Consequently, for some $\mathcal{EL}$ TBoxes admissible TIPs based on finite learning sets might not exist, and so finite learnability cannot be achieved in general.

One trivial way to tame this behavior is to "finitize" $\mathcal{T}^{\not\models}$ by delimiting the entire space of possible TBox axioms to a pre-defined, finite set. This can be achieved, for instance, by restricting the permitted depth of complex concepts or generally setting some a priori bound on the size of axioms. Such ad hoc solutions, though likely efficient in practice, are not very elegant. As a more interesting alternative, we are able to show that there exist at least two languages between $\mathcal{L}^{\sqcap}$ and $\mathcal{EL}$, namely $\mathcal{EL}^{lhs}$ and $\mathcal{EL}^{rhs}$, for which finite learning sets are always guaranteed to exist, regardless of the fact that they permit infinitely many concept inclusions. In fact, we demonstrate that in both cases such learning sets might well consist of exactly one exemplary finite model.

We adopt the technique of so-called *types*, known from the area of modal logics [12]. Types are finite abstractions of possible individuals in the interpretation

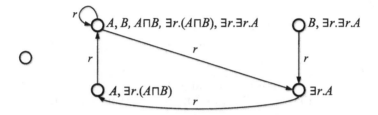

Fig. 3. A finite learning set for an $\mathcal{EL}^{\mathrm{rhs}}$ TBox $\{A \sqsubseteq \exists r.(A \sqcap B), B \sqsubseteq \exists r.\exists r.A\}$. The figure includes type contents (in grey), as defined in the proof of Theorem 2.

domain, out of which arbitrary models can be constructed. Let $con(\mathcal{T})$ be the set of all concepts (and all their subconcepts) occurring in $\mathcal{T}$. A *type* over $\mathcal{T}$ is a set $t \subseteq con(\mathcal{T})$, such that $C \sqcap D \in t$ iff $C \in t$ and $D \in t$, for every $C \sqcap D \in con(\mathcal{T})$. A type t is *saturated* for $\mathcal{T}$ iff for every $C \sqsubseteq D \in \mathcal{T}$, if $C \in t$ then $D \in t$. For any $S \subseteq con(\mathcal{T})$, we write t_S to denote the smallest saturated type containing S. It is easy to see, that t_S must be unique for $\mathcal{EL}$.

The next theorem addresses the case of $\mathcal{EL}^{\mathrm{rhs}}$. Figure 3 illustrates a finite learning set for a sample $\mathcal{EL}^{\mathrm{rhs}}$ TBox, following the construction in the proof.

Theorem 2 (Finite learning sets in $\mathcal{EL}^{\mathrm{rhs}}$). *Let $\mathcal{T}$ be a TBox in $\mathcal{EL}^{\mathrm{rhs}}$. There exists a finite set S such that $(\mathcal{T}, S)$ is admissible.*

Proof sketch. Let Θ be the smallest set of types satisfying the following conditions:

- $t_S \in \Theta$, for every $S \subseteq N_C$ and for $S = \{\top\}$,
- if $t \in \Theta$ then $t_{\{C\}} \in \Theta$, for every $\exists r.C \in t$.

We define the interpretation $\mathcal{I} = (\Delta^{\mathcal{I}}, \cdot^{\mathcal{I}})$ as follows:

- $\Delta^{\mathcal{I}} := \Theta$,
- $t \in A^{\mathcal{I}}$ iff $A \in t$, for every $t \in \Theta$ and $A \in N_C$,
- $(t, t_{\{C\}}) \in r^{\mathcal{I}}$, for every $t \in \Theta$, whenever $\exists r.C \in t$.

Then $S = \{\mathcal{I}\}$ is a finite learning set such that $(\mathcal{T}, S)$ is admissible. ❑

A similar, though somewhat more complex construction demonstrates the existence of finite learning sets in $\mathcal{EL}^{\mathrm{lhs}}$. Again, we illustrate the approach with an example in Fig. 4.

Theorem 3 (Finite learning sets in $\mathcal{EL}^{\mathrm{lhs}}$). *Let $\mathcal{T}$ be a TBox in $\mathcal{EL}^{\mathrm{lhs}}$. There exists a finite set S such that $(\mathcal{T}, S)$ is admissible.*

Proof Sketch. Let Θ be the set of all saturated types over $\mathcal{T}$, and Θ^* be its subset obtained by iteratively eliminating all those types t that violate the following condition: for every $r \in N_R$ and every existential restriction $\exists r.C \in t$ there is $u \in \Theta^*$ such that:

$-\ C \in u,$
$-\ $ for every $\exists r.D \in con(\mathcal{T})$, if $D \in u$ then $\exists r.D \in t$.

Further, we define the interpretation $\mathcal{I} = (\Delta^{\mathcal{I}}, \cdot^{\mathcal{I}})$ as follows:

$-\ \Delta^{\mathcal{I}} := \Theta^*$,
$-\ t \in A^{\mathcal{I}}$ iff $A \in S_t$, for every $t \in \Theta^*$ and $A \in N_C$,
$-\ (t, u) \in r^{\mathcal{I}}$ iff for every $\exists r.C \in con(\mathcal{T})$, if $C \in u$ then $\exists r.C \in t$.

Then $\mathcal{S} = \{\mathcal{I}\}$ is a finite learning set such that $(\mathcal{T}, \mathcal{S})$ is admissible. $\qquad \Box$

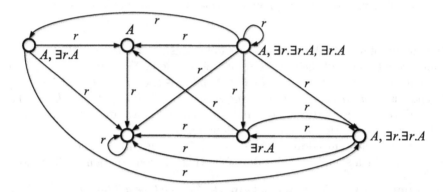

Fig. 4. A finite learning set for an $\mathcal{EL}^{\mathrm{lhs}}$ TBox $\{\exists r.\exists r.A \sqsubseteq A\}$. The figure includes type contents (in grey), as defined in the proof of Theorem 3.

5   Learning Algorithms

In this section, we devise two learning algorithms for admissible TIPs with finite learning sets that correctly identify 1) $\mathcal{EL}^{\mathrm{lhs}}$ and $\mathcal{EL}^{\mathrm{rhs}}$ TBoxes using an equivalence oracle, and 2) $\mathcal{EL}^{\mathrm{rhs}}$ TBoxes without such an oracle, i.e., in a fully unsupervised manner.

Since the set $\mathcal{T}^{\not\models} = \{C \sqsubseteq D$ in $\mathcal{L} \mid \mathcal{T} \not\models C \sqsubseteq D\}$ can be in general infinite, our starting observation is that a learner cannot effectively eliminate concept inclusions from $\mathcal{T}^{\not\models}$ using a straightforward enumeration, thus arriving at the target TBox $\mathcal{T}$. The only feasible strategy is to try to identify the "good" candidate axioms to be included in $\mathcal{T}$, and possibly apply the elimination strategy only to finitely many incorrect guesses. One generic procedure to employ such heuristic, which we define as Algorithm 1, attempts to construct the hypothesis by extending it with consecutive axioms of systematically growing size that are satisfied by the learning set. There, by $\ell(C \sqsubseteq D)$ we denote the size of the axiom $C \sqsubseteq D$ measured in the total number of symbols used for expressing this axiom. At each step the algorithm makes use of a simple equivalence oracle, which informs whether the currently considered hypothesis is already equivalent to the learning target (in that case the identification succeeds) or whether some axioms are still missing. Theorem 4 demonstrates the correctness of this approach.

Algorithm 1. Learning $\mathcal{EL}^{\mathrm{rhs}}/\mathcal{EL}^{\mathrm{lhs}}$ TBoxes on finite inputs.

Input: a TIP $(\mathcal{T}, \mathcal{S})$
Output: a hypothesis TBox $\mathcal{H}$
1: $n := 2$
2: $\mathcal{H}_n := \emptyset$
3: **while** '$\mathcal{H}_n \equiv \mathcal{T}$'? is 'NO' (*equivalence oracle querying*) **do**
4:    $n := n + 1$
5:    $\mathsf{Cand}_n := \{C \sqsubseteq D \in \mathcal{EL}^{\mathrm{rhs}}/\mathcal{EL}^{\mathrm{lhs}} \mid \ell(C \sqsubseteq D) = n\}$
6:    $\mathsf{Accept}_n := \{C \sqsubseteq D \in \mathsf{Cand}_n \mid \mathcal{S} \models C \sqsubseteq D\}$
7:    $\mathcal{H}_n := \mathcal{H}_{n-1} \cup \mathsf{Accept}_n$
8: **end while**
9: **return** $\mathcal{H}_n$

Theorem 4 (Correct identification in $\mathcal{EL}^{\mathrm{rhs}}/\mathcal{EL}^{\mathrm{lhs}}$). *Let $(\mathcal{T}, \mathcal{S})$ be an admissible TIP for $\mathcal{T}$ in $\mathcal{EL}^{\mathrm{rhs}}/\mathcal{EL}^{\mathrm{lhs}}$. Then the hypothesis TBox $\mathcal{H}$ generated by Algorithm 1 is equivalent to $\mathcal{T}$.*

Obviously the use of the oracle is essential to warrant termination of the algorithm. It is not difficult to see that without it, the algorithm must still converge on the correct TBox for some $n \in \mathbb{N}$, and consequently settle on it, i.e., $\mathcal{H}_m \equiv \mathcal{H}_n$ for every $m \geq n$. However, at no point of time can it guarantee that the convergence has been already achieved, and so it can only warrant learnability in the limit. This result is therefore not entirely satisfactory considering we aim at finite learnability from data in the unsupervised setting.

A major positive result, on the contrary, can be delivered for the case of $\mathcal{EL}^{\mathrm{rhs}}$, for which we devise an effective learning algorithm making no reference to any oracle. It turns out that in $\mathcal{EL}^{\mathrm{rhs}}$ the "good" candidate axioms can be directly extracted from the learning set, thus granting a proper unsupervised learning method. The essential insight is provided by Lemma 1, presented in the previous section. Given any $\mathcal{L}^{\sqcap}$ concept C such that $\mathcal{S}(C) \neq \emptyset$ we are able to identify a tree-shaped minimal model of C w.r.t. $\mathcal{T}$. Effectively, it suffices to retrieve only the initial part of this model, discarding its infinitely recurrent (cyclic) subtrees. Such an initial model $\mathcal{I}_{\mathrm{init}}$ is constructed by Algorithm 2. The algorithm performs simultaneous unravelling of all models in $\mathcal{S}(C)$, while on the way, computing intersections of visited combinations of individuals, which are subsequently added to the model under construction. Whenever the same combination of individuals is about to be visited for the third time on the same branch it is skipped, as the cycle is evidently detected and further unravelling is unnecessary. The covering concept $C_{\mathcal{I}_{\mathrm{init}}}$ for the resulting interpretation $\mathcal{I}_{\mathrm{init}}$ is then included in the hypothesis within the axiom $C \sqsubseteq C_{\mathcal{I}_{\mathrm{init}}}$. Meanwhile, all $\mathcal{L}^{\sqcap}$ concepts C such that $\mathcal{S}(C) = \emptyset$ are ensured to entail every $\mathcal{EL}$ concept, as implied by the admissibility condition. The contents of the hypothesis TBox are formally specified in Definition 5.

Algorithm 2. Computing the initial part of the minimal model $\bigcap \mathcal{S}(C)$

Input: the set $\mathcal{S}(C) = \{(\mathcal{I}_i, d_i)\}_{0 \leq i \leq n}$, for some $n \in \mathbb{N}$
Output: a finite tree-shaped interpretation $(\mathcal{J}, d)$, where $\mathcal{J} = (\Delta^{\mathcal{J}}, \cdot^{\mathcal{J}})$
 1: $\Delta^{\mathcal{J}} := \{f(d_0, \ldots, d_n)\}$, for a "fresh" function symbol f
 2: $A^{\mathcal{J}} := \emptyset$, for every $A \in N_C$
 3: $r^{\mathcal{J}} := \emptyset$, for every $r \in N_R$
 4: **for** every $f(d_0, \ldots, d_n) \in \Delta^{\mathcal{J}}$, $(e_0, \ldots, e_n) \in \Delta^{\mathcal{I}_0} \times \ldots \times \Delta^{\mathcal{I}_n}$, $r \in N_R$ **do**
 5:     **if** $(d_i, e_i) \in r^{\mathcal{I}_i}$ for every $0 \leq i \leq n$ **and** there exists no function symbol g
        such that $g(e_0, \ldots, e_n)$ is an ancestor of $f(d_0, \ldots, d_n)$ in $\mathcal{J}$ **then**
 6:         $\Delta^{\mathcal{J}} := \Delta^{\mathcal{J}} \cup \{g(e_0, \ldots, e_n)\}$, for a "fresh" function symbol g
 7:         $r^{\mathcal{J}} := r^{\mathcal{J}} \cup \{(f(d_0, \ldots, d_n), g(e_0, \ldots, e_n))\}$
 8:     **end if**
 9: **end for**
10: **for** every $f(d_0, \ldots, d_n) \in \Delta^{\mathcal{J}}$, $A \in N_C$ **do**
11:     **if** $d_i \in A^{\mathcal{I}_i}$ for every $0 \leq i \leq n$ **then**
12:         $A^{\mathcal{J}} := A^{\mathcal{J}} \cup \{f(d_0, \ldots, d_n)\}$
13:     **end if**
14: **end for**
15: **return** $(\mathcal{J}, f(d_0, \ldots, d_n))$, where $f(d_0, \ldots, d_n)$ is the root of $\mathcal{J}$, created at
    step 1.

Definition 5 ($\mathcal{EL}^{\mathrm{rhs}}$ hypothesis TBox). *Let $(\mathcal{T}, \mathcal{S})$ be an admissible TIP for $\mathcal{T}$ in $\mathcal{EL}^{\mathrm{rhs}}$ over the vocabulary $\Sigma_{\mathcal{T}}$. The hypothesis TBox $\mathcal{H}$ is the set consisting of all the following axioms:*

- *$C \sqsubseteq C_{\mathcal{I}_{init}}$ for every $\mathcal{L}^{\sqcap}$ concept C such that $\mathcal{S}(C) \neq \emptyset$, where $C_{\mathcal{I}_{init}}$ is the covering concept for the interpretation $(\mathcal{I}_{init}, d)$ generated by Algorithm 2 on $\mathcal{S}(C)$;*
- *$C \sqsubseteq \bigcap_{r \in N_R} \exists r. \bigcap N_C$ for every $\mathcal{L}^{\sqcap}$ concept C such that $\mathcal{S}(C) = \emptyset$.*

To better illustrate the learning procedure, we consider a simple TIP consisting of an $\mathcal{EL}^{\mathrm{rhs}}$ TBox $\mathcal{T} = \{A \sqsubseteq \exists r.(A \sqcap B)\}$ and a finite learning set $\mathcal{S} = \{\mathcal{I}\}$, with $\mathcal{I}$ as depicted in Fig. 5. The assumed vocabulary containing two concept names — A and B — induces four distinct $\mathcal{L}^{\sqcap}$ concepts, namely: $\top$, A, B and $A \sqcap B$. For every such concept C we identify the corresponding set of its all pointed models $\mathcal{S}(C)$ contained in the learning set. For instance, $\mathcal{S}(A) = \{(\mathcal{I}, e_2), (\mathcal{I}, e_3)\}$. Further, we use Algorithm 2 to compute the initial part of the minimal model $\bigcap \mathcal{S}(C)$, as illustrated in Fig. 6. Finally, based on these models we formulate the hypothesis TBox, as specified in Definition 5: $\mathcal{H} = \{\top \sqsubseteq \top, A \sqsubseteq \exists r.(A \sqcap B \sqcap \exists r.(A \sqcap B)), B \sqsubseteq B, A \sqcap B \sqsubseteq \exists r.(A \sqcap B)\}$. It is not difficult to verify that $\mathcal{H} \equiv \mathcal{T}$.

The correctness of the learning procedure is demonstrated in the following theorem.

Theorem 5 (Correct identification in $\mathcal{EL}^{\mathrm{rhs}}$). *Let $(\mathcal{T}, \mathcal{S})$ be an admissible TIP for $\mathcal{T}$ in $\mathcal{EL}^{rhs}$. Then the hypothesis TBox $\mathcal{H}$ for $\mathcal{S}$ is equivalent to $\mathcal{T}$.*

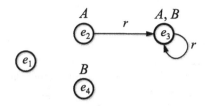

Fig. 5. A finite learning set for an $\mathcal{EL}^{\text{rhs}}$ TBox $\{A \sqsubseteq \exists r.(A \sqcap B)\}$.

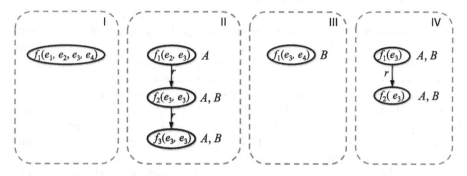

Fig. 6. The initial parts of the minimal models $\bigcap \mathcal{S}(C)$ computed with Algorithm 2 over the learning set in Fig. 5, where (I) $C = \top$, (II) $C = A$, (III) $C = B$, (IV) $C = A \sqcap B$.

Proof sketch. Let C be a concept in $\mathcal{L}^{\sqcap}$ such that $\mathcal{S}(C) \neq \emptyset$. By Lemma 1, the intersection $(\mathcal{J}, d) = \bigcap \mathcal{S}(C)$ is a minimal model of C w.r.t. $\mathcal{T}$. Without loss of generality, we assume that $(\mathcal{J}, d)$ is a tree-shaped model. We also note, that every $\mathcal{EL}$ concept C induces a syntactic tree, which corresponds directly to a minimal model of C. It is not difficult to see that Algorithm 2 indeed produces an initial part $(\mathcal{J}_{\text{init}}, d)$ of $(\mathcal{J}, d)$. By reconstructing the concept C_{init} from we in fact identify all minimal (i.e., necessary) consequences of C w.r.t. $\mathcal{T}$. However, certain infinite subtrees of $(\mathcal{J}, d)$ are omitted in $(\mathcal{J}_{\text{init}}, d)$. This happens due to the condition at step 5 of Algorithm 2, which terminates the construction of certain branches whenever a cycle is detected. In the rest of the proof, we show that the covering concept $C_{\mathcal{J}_{\text{init}}}$ has the same minimal model w.r.t. $\mathcal{H}$ as C has w.r.t. $\mathcal{T}$. Since this is demonstrated to hold for every $\mathcal{L}^{\sqcap}$ concept C, we can conclude that $\mathcal{H} \equiv \mathcal{T}$. □

The learning algorithm runs in double exponential time in the worst case and generates TBoxes of double exponential size in the size of $\mathcal{S}$. This follows from the fact that the tree-shaped interpretations generated by Algorithm 2 might be of depth exponential in the number of individuals occurring in $\mathcal{S}$ and have exponential branching factor. Importantly, however, there might exist solutions far closer to being optimal which we have not as far investigated.

It is our strong conjecture, which we leave as an open problem, that a similar learning strategy should also be applicable in the context of $\mathcal{EL}^{\text{lhs}}$.

6   Related Work

Ontology learning is an interdisciplinary research field drawing on techniques from Formal Concept Analysis [13,14], Natural Language Processing [15,16] and machine learning [6,16,17], to name a few. One classification of ontology learning techniques distinguishes between investigations of exact learnability, and approaches incorporating probabilistic, vague or fuzzy reasoning [18,19]. Another classification is at the level at which learning takes place [15] — does the problem address learning of concepts, concept hierarchies, logical theories or rules? Lehmann and Völker [17] distinguishes between four types of ontology learning: learning from text, data mining, concept learning and crowdsourcing.

In this landscape, the present paper is on exact learnability and, within this framework, addresses the problem of learning logical theories. That is, we address the problem at the level of relationships between concepts, positing a logical theory, rather than at the concept level, learning concept descriptions [20–24]. Furthermore, the target theory is identified from interpretations, and is hence related to various contributions on learnability of different types of formal structures from data, e.g.: first-order theories from facts [10], finite automata descriptions from observations [25], and logic programs from interpretations [5,6].

The model for exact learning of DL TBoxes which offers the most direct comparison to ours was introduced recently by Konev, et al. [8], and follows on prior research by the same authors based on Angluin's model of learning from entailment [7,26]. In their learning framework for learning from data retrieval queries, the learner identifies the TBox by posing two types of queries to an oracle: membership queries of the form "$(\mathcal{T}, \mathcal{A}) \models q$?", where $\mathcal{A}$ is a given ABox and q is a query, and equivalence queries of the form "Does the hypothesis ontology $\mathcal{H}$ entail the target ontology $\mathcal{T}$?". The authors study polynomial learnability in fragments of $\mathcal{EL}$ and DL-Lite, and for queries ranging from atomic to conjunctive queries.

Essentially, given a finite learning set in an admissible TIP, a learner from interpretations can autonomously answer arbitrary membership queries, thus effectively simulating the membership oracle. However, it does not have by default access to an equivalence oracle. Once such an oracle is included, as in Algorithm 1, the learning power of both learners becomes comparable for the languages investigated in the present paper. In this sense, our Theorem 4 should be also indirectly derivable from the results by Konev et al. However, our stronger result for $\mathcal{EL}^{\mathrm{rhs}}$ in Theorem 5 demonstrates that, at least in some cases, the learner from interpretations is able to succeed without employing any oracle. While learning from ABoxes and query answers makes sense in a semi-automated learning environment, learning from interpretations is in our view a more appropriate model in the context of fully autonomous learning.

7   Conclusions and Outlook

In this paper, we have delivered initial results on finite learnability of DL TBoxes from interpretations. We believe that this direction shows promise in establishing formal foundations for the task of ontology learning from data. Some immediate problems that are left open with this work concern finite learnability of $\mathcal{EL}^{\text{lhs}}$ TBoxes in an unsupervised setting, and possibly of other lightweight fragments of DLs. Another set of very interesting research questions should deal, in our view, with the possibility of formulating alternative conditions on the learning sets and the corresponding learnability guarantees they would imply in different DL languages. In particular, some limited use of closed-world operator over the learning sets might allow to relax the practically restrictive admissibility condition. Finally, the development of practical learning algorithms, possibly building on existing inductive logic programming methods, is an obvious area to welcome further research efforts.

References

1. Maedche, A., Staab, S.: Ontology learning. In: Staab, S., Studer, R. (eds.) Handbook on Ontologies, pp. 173–189. Springer, New York (2004)
2. Hoekstra, R.: The knowledge reengineering bottleneck. J. Semant. Web **1**(1,2), 111–115 (2010)
3. Baader, F., Calvanese, D., Mcguinness, D.L., Nardi, D., Patel-Schneider, P.F.: The Description Logic Handbook: Theory, Implementation, and Applications. Cambridge University Press, New York (2003)
4. Baader, F., Brandt, S., Lutz, C.: Pushing the $\mathcal{EL}$ envelope. In: Proceedings of IJCAI-05 (2005)
5. De Raedt, L., Lavrač, N.: The many faces of inductive logic programming. In: Komorowski, J., Raś, Z.W. (eds.) ISMIS 1993. LNCS, vol. 689, pp. 435–449. Springer, Heidelberg (1993)
6. De Raedt, L.: First order jk-clausal theories are PAC-learnable. Artif. Intell. **70**, 375–392 (1994)
7. Konev, B., Lutz, C., Ozaki, A., Wolter, F.: Exact learning of lightweight description logic ontologies. In: Proceedings of Principles of Knowledge Representation and Reasoning (KR-14) (2014)
8. Konev, B., Lutz, C., Wolter, F.: Exact learning of TBoxes in $\mathcal{EL}$ and DL-Lite. In: Proceedings of the 28th International Workshop on Description Logics (2015)
9. Lutz, C., Piro, R., Wolter, F.: Enriching $\mathcal{EL}$-concepts with greatest fixpoints. In: Proceedings of the 19th European Conference on Artificial Intelligence (ECAI 2010), pp. 41–46. IOS Press (2010)
10. Shapiro, E.Y.: Inductive inference of theories from facts. In: Computational Logic: Essays in Honor of Alan Robinson (1991). MIT Press (1981)
11. Klarman, S., Britz, K.: Ontology learning from interpretations in lightweight description logics. Technical report, CSIR Centre for Artificial Intelligence Research, South Africa (2015). http://klarman.synthasite.com/resources/KlaBri-ILP15.pdf
12. Pratt, V.: Models of program logics. In: Proceedings of Foundations of Computer Science (FOCS 1979) (1979)
13. Baader, F., Ganter, B., Sertkaya, B., Sattler, U.: Completing description logic knowledge bases using formal concept analysis. In: Proceedings of the 20th International Joint Conference on Artificial Intelligence (IJCAI-07) (2007)
14. Distel, F.: Learning description logic knowledge bases from data using methods from formal concept analysis. Ph.D. Thesis, TU Dresden (2011)

15. Buitelaar, P., Cimeano, P., Magnini, F. (eds.): Ontology Learning from Text: Methods, Evaluation and Applications. IOS Press, Amsterdam (2005)
16. Cimeano, P., Mädche, A., Staab, S., Völker, J.: Ontology learning. In: Staab, S., Studer, R. (eds.) Handbook on Ontologies. Springer, New York (2009)
17. Lehmann, J., Völker, J. (eds.): Perspectives on Ontology Learning. IOS Press, Amsterdam (2014)
18. Cohen, W., Hirsh, H.: The learnability of description logics with equality constraints. Mach. Learn. **17**(2–3), 169–199 (1994)
19. Lisi, F.A., Straccia, U.: A FOIL-like method for learning under incompleteness and vagueness. In: Zaverucha, G., Santos Costa, V., Paes, A. (eds.) ILP 2013. LNCS, vol. 8812, pp. 123–139. Springer, Heidelberg (2014)
20. Badea, L., Nienhuys-Cheng, S.-H.: A refinement operator for description logics. In: Cussens, J., Frisch, A.M. (eds.) ILP 2000. LNCS (LNAI), vol. 1866, pp. 40–59. Springer, Heidelberg (2000)
21. Lehmann, J., Hitzler, P.: A refinement operator based learning algorithm for the $\mathcal{ALC}$ description logic. In: Blockeel, H., Ramon, J., Shavlik, J., Tadepalli, P. (eds.) ILP 2007. LNCS (LNAI), vol. 4894, pp. 147–160. Springer, Heidelberg (2008)
22. Fanizzi, N., d'Amato, C., Esposito, F.: DL-FOIL concept learning in description logics. In: Železný, F., Lavrač, N. (eds.) ILP 2008. LNCS (LNAI), vol. 5194, pp. 107–121. Springer, Heidelberg (2008)
23. Cohen, W.W., Hirsh, H.: Learning the classic description logic: Theoretical and experimental results. In: Proceedings of Principles of Knowledge Representation and Reasoning (KR 1994) (1994)
24. Chitsaz, M., Wang, K., Blumenstein, M., Qi, G.: Concept learning for $\mathcal{EL}++$ by refinement and reinforcement. In: Anthony, P., Ishizuka, M., Lukose, D. (eds.) PRICAI 2012. LNCS, vol. 7458, pp. 15–26. Springer, Heidelberg (2012)
25. Pitt, L.: Inductive inference, DFAs, and computational complexity. In: Jantke, K.P. (ed.) All 1989. LNCS, vol. 397, pp. 18–44. Springer, Heidelberg (1989)
26. Angluin, D.: Queries and concept learning. Mach. Learn. **2**(4), 319–342 (1988)

Constructing Markov Logic Networks from First-Order Default Rules

Ondřej Kuželka[1]([✉]), Jesse Davis[2], and Steven Schockaert[1]

[1] School of Computer Science & Informatics, Cardiff University, Cardiff, UK
{KuzelkaO,SchockaertS1}@cardiff.ac.uk
[2] Department of Computer Science, Katholieke Universiteit Leuven, Leuven, Belgium
jesse.davis@cs.kuleuven.be

Abstract. Expert knowledge can often be represented using default rules of the form "if A then typically B". In a probabilistic framework, such default rules can be seen as constraints on what should be derivable by MAP-inference. We exploit this idea for constructing a Markov logic network $\mathcal{M}$ from a set of first-order default rules D, such that MAP inference from $\mathcal{M}$ exactly corresponds to default reasoning from D, where we view first-order default rules as templates for the construction of propositional default rules. In particular, to construct appropriate Markov logic networks, we lift three standard methods for default reasoning. The resulting Markov logic networks could then be refined based on available training data. Our method thus offers a convenient way of using expert knowledge for constraining or guiding the process of learning Markov logic networks.

1 Introduction

Markov logic is a popular framework for statistical relational learning [20]. Formulas in Markov logic essentially correspond to weighted first-order formulas, which act as soft constraints on possible worlds. In current applications, the weights are typically learned from data, while the first-order formulas are either hand crafted or obtained using standard rule learning methods.

The fact that a domain expert could manually specify (some of) the formulas, or could inspect learned formulas, is an important strength of Markov logic. Unfortunately, the weights associated with these formulas do not have an easily interpretable meaning. This limits the potential of Markov logic, as it means that domain experts cannot offer much guidance in terms of how the weights should be set (e.g. in applications with little or no training data) or cannot verify the quality of learned weights (e.g. in applications where the quality of the training data is in doubt). Often, however, Markov logic networks (MLN) are not used for evaluating probabilities but for finding the most likely truth assignment of unobserved variables, given the available evidence, i.e. for *maximum a posteriori* (MAP) reasoning. In such cases, the precise values of the weights are only relevant inasmuch as they influence the result of MAP queries. In this setting, we can instead ask for constraints on how MAP reasoning should behave as opposed to

© Springer International Publishing Switzerland 2016
K. Inoue et al. (Eds.): ILP 2015, LNAI 9575, pp. 91–105, 2016.
DOI: 10.1007/978-3-319-40566-7_7

asking a domain expert to specify weights. For example, the expert could specify constraints such as "if all we know is that x is a bird, then using MAP reasoning we should be able to conclude that x can fly", which is in agreement with the semantics of the default rule "birds can typically fly" in System P [2,13].

Thus, a domain expert could be involved in the process of learning an MLN by providing a set of defaults, which are interpreted as constraints on the ranking of possible worlds induced by the MLN. Taking this idea one step further, in this paper, we show how a specific MLN can be constructed from the default rules provided by the expert. Constructing this MLN requires us to select a specific probability distribution that is compatible with the defaults. This selection problem is closely related to the problem of defining the closure of a set of defaults, which has been widely studied in the field of non-monotonic reasoning [9,10,14]. In particular, several proposals to define this closure are based on constructing a specific probability distribution [4,10]. As we will show, it is possible to lift these approaches and thus obtain an efficient and principled way to construct MLNs that are compatible with a given set of defaults.

To date, the use of expert knowledge for guiding or even replacing weight learning has only received limited attention. One exception is [17], which constructs an MLN based on (potentially inconsistent) conditional probabilities provided by domain experts. While this can be useful in some applications, it relies on the ability of experts to provide meaningful probability estimates. However, humans are notoriously poor at judging likelihood. For example, properties that are common among the typical elements of a class of objects are often assumed to be likely in general [22]. Moreover, experts may be able to specify which properties are most likely to hold, in a given context, without being able to quantify their likelihood. In such situations, our default-rule-based approach would be more natural than approaches that force experts to estimate probabilities. On the other hand, our approach will only provide meaningful results for MAP queries: numerical input will be difficult to avoid if we want the constructed MLN to produce satisfactory conditional probability estimates.

This paper is structured as follows. The next section recalls some preliminaries from Markov logic and the non-monotonic reasoning literature. Then in Sect. 3 we show how three well-known approaches to non-monotonic reasoning can be implemented as MAP inference in a particular MLN. By lifting the constructions from Sect. 3, in Sect. 4 we show how MLNs can be constructed whose MAP-consequences are compatible with a given set of first-order default rules. Finally, Sect. 5 evaluates the performance of the resulting MLNs in a standard classification task.

2   Background

2.1   Markov Logic Networks

A Markov logic network (MLN) [20] is a set of weighted formulas (F, w_F), where F is a classical first-order formula and w_F is a real number, intuitively reflecting a penalty that is applied to possible worlds that violate F. We will sometimes also use the notation $w_F : F$ to denote the formula (F, w_F). Given a set of constants C, an MLN $\mathcal{M}$ induces the following probability distribution on the set of possible worlds ω:

$$p_{\mathcal{M}}(\omega) = \frac{1}{Z} \exp \left(\sum_{(F, w_F) \in \mathcal{M}} w_F n_F(\omega) \right), \qquad (1)$$

where $n_F(x)$ is the number of true groundings of F in the possible world ω, and $Z = \sum_\omega p(\omega)$ is a normalization constant to ensure that p can be interpreted as a probability distribution. Sometimes, formulas (F, w_F) with $w_F = +\infty$ are considered to represent hard constraints. In such cases, we define $p_{\mathcal{M}}(\omega) = 0$ for all possible worlds that do not satisfy all of the hard constraints, and only formulas with a real-valued weight are considered in (1) for the possible worlds that do.

The main inference task for MLNs which we will consider is full MAP inference. Given a set of ground literals (the evidence), MAP inferences aims to compute the most probable configuration of all unobserved variables (the queries). Standard approaches for performing MAP inference include a strategy based on MaxWalkSAT [20] and a cutting plane based strategy [16,21]. Given a set of ground formulas E, we write $\max(\mathcal{M}, E)$ for the set of most probable worlds of the MLN that satisfy E. We will also consider the following inference relation, initially proposed for penalty logic in [7]:

$$(\mathcal{M}, E) \vdash_{MAP} \alpha \quad \text{iff} \quad \forall \omega \in \max(\mathcal{M}, E) : \omega \models \alpha \qquad (2)$$

with $\mathcal{M}$ an MLN, α a ground formula and E a set of ground formulas. Note that $(\mathcal{M}, E) \vdash_{MAP} \alpha$ means that the formula α is satisfied in all the most probable worlds which are compatible with the available evidence.

2.2 Reasoning About Default Rules in System P

A variety of approaches have been proposed to reason about default rules of the form "if α then typically β holds", which we will denote as $\alpha \mathrel{|\!\sim} \beta$. Most approaches are based on the idea of defining a preference order over possible worlds and insisting that β is true in the most preferred (i.e. the most normal) of the worlds in which α is true [3,9,13,18,19]. The axioms of System P [13] capture a set of desirable properties for an inference relation for default rules:

Reflexivity. $\alpha \mathrel{|\!\sim} \alpha$
Left logical equivalence. If $\alpha \equiv \alpha'$ and $\alpha \mathrel{|\!\sim} \beta$ then $\alpha' \mathrel{|\!\sim} \beta$
Right weakening. If $\beta \models \beta'$ and $\alpha \mathrel{|\!\sim} \beta$ then $\alpha \mathrel{|\!\sim} \beta'$
OR. If $\alpha \mathrel{|\!\sim} \gamma$ and $\beta \mathrel{|\!\sim} \gamma$ then $\alpha \vee \beta \mathrel{|\!\sim} \gamma$
Cautious monotonicity. If $\alpha \mathrel{|\!\sim} \beta$ and $\alpha \mathrel{|\!\sim} \gamma$ then $\alpha \wedge \beta \mathrel{|\!\sim} \gamma$
CUT. If $\alpha \wedge \beta \mathrel{|\!\sim} \gamma$ and $\alpha \mathrel{|\!\sim} \beta$ then $\alpha \mathrel{|\!\sim} \gamma$

where $\alpha \equiv \alpha'$ and $\beta \models \beta'$ refer to equivalence and entailment from classical logic. Note that applying the axioms of System P to a set of defaults $\Delta = \{\alpha_1 \mathbin{\mid\!\sim} \beta_1, ..., \alpha_n \mathbin{\mid\!\sim} \beta_n\}$ corresponds to a form of monotonic reasoning about defaults. However, as the set of consequences that can be obtained in this way is limited, it is common to consider a non-monotonic inference relation whose set of consequences is closed under the axioms of System P as well as the following property:

Rational monotonicity If $\alpha \mathbin{\mid\!\sim} \beta$ and we cannot derive $\alpha \mathbin{\mid\!\sim} \neg\gamma$ then $\alpha \wedge \gamma \mathbin{\mid\!\sim} \beta$.

In this paper, we will consider three such inference relations: the rational closure [19], the lexicographic closure and the maximum entropy closure. A default $\alpha \mathbin{\mid\!\sim} \beta$ is tolerated by a set of defaults $\gamma_1 \mathbin{\mid\!\sim} \delta_1, ..., \gamma_m \mathbin{\mid\!\sim} \delta_m$ if the classical formula $\alpha \wedge \beta \wedge \bigwedge_i (\neg\gamma_i \vee \delta_i)$ is consistent. The rational closure is based on a stratification $\Delta_1, ..., \Delta_k$ of Δ, where each Δ_j contains all defaults $\alpha \mathbin{\mid\!\sim} \beta$ from $\Delta \setminus (\Delta_1 \cup ...\Delta_{j-1})$ which are tolerated by $\Delta \setminus (\Delta_1 \cup ... \cup \Delta_{j-1})$. It can be shown that such a stratification always exists when Δ satisfies some natural consistency properties (see [19] for details). Intuitively, Δ_1 contains the most general default rules, Δ_2 contains exceptions to the rules in Δ_1, Δ_3 contains exceptions to the rules in Δ_2, etc. This stratification is known as the Z-ordering. Let j be the smallest index for which $\Delta_\alpha^{rat} = \{\neg\alpha \vee \beta | \alpha \mathbin{\mid\!\sim} \beta \in \Delta_j \cup ... \cup \Delta_k\} \cup \{\alpha\}$ is consistent. Then $\alpha \mathbin{\mid\!\sim} \beta$ is in the rational closure of Δ if $\Delta_\alpha^{rat} \models \beta$. When a set of hard rules Γ needs to be enforced, the Z-ordering can be generalized as follows [3]. Each set Δ_j then contains those defaults $\alpha \mathbin{\mid\!\sim} \beta$ for which $\Gamma \cup \{\alpha \wedge \beta\} \cup \{\neg\alpha_i \vee \beta_i : (\alpha_i \mathbin{\mid\!\sim} \beta_i) \in \Delta \setminus (\Delta_1 \cup ... \cup \Delta_{j-1})\}$. Finally, we define $\Delta_k = \Gamma$, where $\Delta_1, ..., \Delta_{k-1}$ is the stratification of Δ that was obtained.

The rational closure encodes the intuition that in case of conflict, specific rules should have priority over more generic ones. However, it requires us to ignore all the defaults in $\Delta_1 \cup ... \cup \Delta_{j-1}$, even defaults which are intuitively unrelated to this conflict. The lexicographic closure [1] addresses this issue as follows. For a propositional interpretation ω, we write $sat(\omega, \Delta_j)$ for the number of defaults satisfied by ω, i.e. $sat(\omega, \Delta_j) = |\{\alpha \mathbin{\mid\!\sim} \beta : (\alpha \mathbin{\mid\!\sim} \beta) \in \Delta_j, \omega \models \neg\alpha \vee \beta\}|$. We say that an interpretation ω_1 is lex-preferred over an interpretation ω_2, written $\omega_1 \prec \omega_2$, if there exists a j such that $sat(\omega_1, \Delta_j) > sat(\omega_2, \Delta_j)$ while $sat(\omega_1, \Delta_i) = sat(\omega_2, \Delta_i)$ for all $i > j$. The default $\alpha \mathbin{\mid\!\sim} \beta$ is in the lexicographic closure of Δ if β is satisfied in all the most lex-preferred models of α, i.e. $\forall \omega \in [\![\alpha]\!] : (\omega \not\models \beta) \Rightarrow \exists \omega' \in [\![\alpha]\!] : \omega' \prec \omega$, where $[\![\alpha]\!]$ is a shorthand for $\{\omega : \omega \models \alpha\}$.

Another approach to default reasoning is based on the principle of maximum entropy [10]. To describe how the maximum-entropy ranking of possible worlds can be computed, we need some additional terminology. A possible world ω is said to falsify a rule $\alpha \mathbin{\mid\!\sim} \beta$ if $\omega \models \alpha \wedge \neg\beta$ and said to verify it if $\omega \models \alpha \wedge \beta$. A set of default rules Δ is said to be a minimal core if for any rule $\alpha \mathbin{\mid\!\sim} \beta$, the set $\{\alpha \mathbin{\mid\!\sim} \neg\beta\} \cup (\Delta \setminus \{\alpha \mathbin{\mid\!\sim} \beta\})$ is a consistent set of default rules, meaning that a Z-ordering of this set exists. Given a minimal core set of defaults Δ, the maximum-entropy ranking is obtained as follows [10]. Let Γ be the set of rules tolerated by Δ. For each rule $r \in \Gamma$, we set $\kappa_{ME}(r) = 1$. While $\Gamma \neq \Delta$ we repeat the following steps. Let Ω be the set of models ω which do not falsify any of the rules in $\Delta \setminus \Gamma$ and verify at least one of these rules. For each model $\omega \in \Omega$, we compute $\kappa_{ME}(\omega) = \sum\{\kappa_{ME}(\alpha \hspace{0.5mm}\vdash\hspace{-2mm}\sim \beta) \colon (\alpha \hspace{0.5mm}\vdash\hspace{-2mm}\sim \beta) \in \Gamma, \omega \models \alpha \wedge \neg\beta\}$. Let ω^* be the model in Ω with minimum rank. Each rule $\alpha \hspace{0.5mm}\vdash\hspace{-2mm}\sim \beta$ that is verified by ω^* is added to Γ, and its rank is computed as $\kappa_{ME}(\alpha \hspace{0.5mm}\vdash\hspace{-2mm}\sim \beta) = 1 + \kappa_{ME}(\omega^*)$.

3 Encoding Ground Default Theories in Markov Logic

It is well-known [14] that any set of defaults Δ which is closed under the axioms of System P and rational monotonicity corresponds to a linear ranking κ of possible worlds, such that $\alpha \hspace{0.5mm}\vdash\hspace{-2mm}\sim \beta$ iff $\kappa(\alpha \wedge \beta) > \kappa(\alpha \wedge \neg\beta)$, where we write $\kappa(\gamma)$ for a formula γ as an abbreviation for $\max\{\kappa(\omega) \colon \omega \models \gamma\}$. Since the ranking κ can be encoded as a probability distribution, and every probability distribution on possible worlds can be represented as an MLN, it is clear that there must exist an MLN $\mathcal{M}$ such that $(\alpha \hspace{0.5mm}\vdash\hspace{-2mm}\sim \beta) \in \Delta$ iff $(\mathcal{M}, \alpha) \vdash_{MAP} \beta$. More generally, for any (i.e. not necessarily closed) set of defaults Δ, there exists an MLN $\mathcal{M}$ such that $(\mathcal{M}, \alpha) \vdash_{MAP} \beta$ iff $\alpha \hspace{0.5mm}\vdash\hspace{-2mm}\sim \beta$ is in the rational closure of Δ, and similar for the lexicographic and maximum-entropy closures. We now show how the MLNs corresponding to these three closures can be constructed.

Transformation 1 (Rational closure). *Let Δ be a set of ground default rules and let Θ be a set of hard constraints (clauses). Let $\Delta_1, ..., \Delta_k$ be the Z-ordering of $\Delta \cup \Theta$. Let the MLN $\mathcal{M}$ be defined as follows: $\bigcup_{i=1}^{k}(\{(\neg a_i \vee \neg\alpha \vee \beta, \infty) \colon \alpha \hspace{0.5mm}\vdash\hspace{-2mm}\sim \beta \in \Delta_i\} \cup \{(a_i, 1)\} \cup \{(\phi, \infty) \colon \phi \in \Theta\}) \cup \bigcup_{i=2}^{k}\{(a_i \vee \neg a_{i-1}, \infty)\}$ where a_i are auxiliary literals. Then $(\mathcal{M}, \alpha) \vdash_{MAP} \beta$ iff $\alpha \hspace{0.5mm}\vdash\hspace{-2mm}\sim \beta$ is in the rational closure of (Δ, Θ).*

Transformation 2 (Lexicographic closure). *Let Δ be a set of ground default rules and let Θ be a set of hard constraints (clauses). Let $\Delta_1, ..., \Delta_k$ be the Z-ordering of $\Delta \cup \Theta$. Let the MLN $\mathcal{M}$ be defined as follows: $\bigcup_{i=1}^{k}\{(\neg\alpha \vee \beta, \lambda_i) \colon \alpha \hspace{0.5mm}\vdash\hspace{-2mm}\sim \beta \in \Delta_i\} \cup \{(\phi, \infty) \colon \phi \in \Theta\}$ where $\lambda_i = 1 + \sum_{j=1}^{i-1}|\Delta_j| \cdot \lambda_j$ for $i > 1$ and $\lambda_1 = 1$. Then $(\mathcal{M}, \alpha) \vdash_{MAP} \beta$ iff $\alpha \hspace{0.5mm}\vdash\hspace{-2mm}\sim \beta$ is in the lexicographic closure of (Δ, Θ).*

Transformation 3 (Maximum-entropy closure). *Let Δ be a set of ground default rules and let Θ be a set of hard constraints (clauses). Let κ be weights of rules corresponding to the maximum-entropy closure of $\Delta \cup \Theta$. Let the MLN $\mathcal{M}$ be defined as follows: $\{(\neg\alpha \vee \beta, \kappa(\alpha \hspace{0.5mm}\vdash\hspace{-2mm}\sim \beta)) \colon \alpha \hspace{0.5mm}\vdash\hspace{-2mm}\sim \beta \in \Delta\} \cup \{(\phi, \infty) \colon \phi \in \Theta\}$. Then $(\mathcal{M}, \alpha) \vdash_{MAP} \beta$ iff $\alpha \hspace{0.5mm}\vdash\hspace{-2mm}\sim \beta$ is in the maximum-entropy closure of (Δ, Θ).*

Example 1. Consider the default rules $\Delta = \{bird \hspace{0.5mm}\vdash\hspace{-2mm}\sim flies, antarctic \wedge bird \hspace{0.5mm}\vdash\hspace{-2mm}\sim \neg flies\}$. Then $\mathcal{M}_1 = \{(\neg a_1 \vee \neg bird \vee flies, \infty), (\neg a_2 \vee \neg antarctic \vee \neg bird \vee \neg flies, \infty), (a_1, 1), (a_2, 1), (a_2 \vee \neg a_1, \infty)\}$ is the result of Transformation 1, and $\mathcal{M}_2 = \{(\neg bird \vee flies, 1), (\neg antarctic \vee \neg bird \vee \neg flies, 2)\}$ is the result of Transformation 2, which in this example coincides with the result of Transformation 3.

4 Encoding Non-ground Default Theories in Markov Logic

While reasoning with default rules has mostly been studied at the propositional level, a few authors have considered first-order default rules [8,12]. Similar as for probabilistic first-order rules [11], two rather different semantics for first-order defaults can be considered. On the one hand, a default such as $P(x) \mathrel{\vert\!\sim} Q(x)$ could mean that the most typical objects that have the property P also have the property Q. On the other hand, this default could also mean that whenever $P(x)$ holds for a given x, in the most normal worlds $Q(x)$ will also hold. In other words, first-order defaults can either model typicality of objects or normality of worlds [8]. In this paper, we will consider the latter interpretation. Given that we only consider finite universes (as is usual in the context of MLNs), we can then see a first order default as a template for propositional defaults. For example $P(x) \mathrel{\vert\!\sim} Q(x)$ can be seen as a compact notation for a set of defaults $\{P(c_1) \mathrel{\vert\!\sim} Q(c_1), ..., P(c_n) \mathrel{\vert\!\sim} Q(c_n)\}$. Note that this approach would not be possible for first-order defaults that model the typicality of objects.

In particular, the first-order default theories we will consider consist of first-order logic formulas (hard rules) and default rules of the form $\alpha \mathrel{\vert\!\sim} \beta$, where α is a conjunction of literals and β is a disjunction of literals. Our approach can be straightforwardly extended to quantified default rules, where the scopes of quantifiers may be the whole default rules, and not just either the antecedent or the consequent of a rule. While this could be of interest, we do not consider this for the ease of presentation.

Definition 1 (Markov logic model of a first-order default theory). *Let (Δ, Θ) be a first-order default theory with Δ the set of default rules and Θ the set of hard rules. A Markov logic network $\mathcal{M}$ is a model of the default logic theory $\Delta \cup \Theta$ if it holds that: (i) $P[X = \omega] = 0$ whenever $\omega \not\models \Theta$, and (ii) for any default rule $\alpha \mathrel{\vert\!\sim} \beta \in \Delta$ and any grounding substitution θ of the unquantified variables of $\alpha \mathrel{\vert\!\sim} \beta$, either $\{\alpha\theta\} \cup \Theta \vdash \bot$ or $(\mathcal{M}, \alpha\theta) \vdash_{MAP} \beta\theta$. We say that (Δ, Θ) is satisfiable if it has at least one model.*

Below we will describe three methods for constructing Markov logic models of first-order default theories, generalizing Transformations 1–3. For convenience, we will use *typed* formulas (nevertheless, we will assume that default rules given on input are not typed for simplicity). For instance, when we have the formula $\alpha = owns(person : X, thing : Y)$ and the set of constants of the type *person* is $\{alice, bob\}$ and the set of constants of the type *thing* is $\{car\}$ then α corresponds to the ground formulas $owns(alice, car)$ and $owns(bob, car)$. In cases where there is only one type, we will not write it explicitly. For a constant or variable x, we write $\tau(x)$ to denote its type. Two formulas F_1 and F_2 (either both conjunctions or both disjunctions of literals) are said to be isomorphic when there is an injective substitution θ of the variables of F_1 such that $F_1\theta \equiv F_2$ (where $\equiv$ denotes logical equivalence). Two default rules $D_1 = \alpha_1 \mathrel{\vert\!\sim} \beta_1$ and $D_2 = \alpha_2 \mathrel{\vert\!\sim} \beta_2$ are said to be isomorphic, denoted $D_1 \approx D_2$, if there exists a substitution θ

of the variables of D_1 such that $\alpha_1\theta \equiv \alpha_2$ and $\beta_1\theta \equiv \beta_2$. Two default theories $\Delta_1 \cup \Theta_1$ and $\Delta_2 \cup \Theta_2$ are said to be isomorphic, denoted by $\Delta_1 \cup \Theta_1 \approx \Delta_2 \cup \Theta_2$, if there is a bijection i from elements of $\Delta_1 \cup \Theta_1$ to elements of $\Delta_2 \cup \Theta_2$ such that for any $F \in \Delta_1 \cup \Theta_1$, $i(F) \approx F$. When j is a permutation of a subset of constants from $\Delta \cup \Theta$ then $j(\Delta \cup \Theta)$ denotes the default theory obtained by replacing any constant c from the subset by its image $j(c)$.

Definition 2 (Interchangeable constants). *Let $\Delta \cup \Theta$ be a non-ground default theory. A set of constants $\mathcal{C}$ is said to be interchangeable in $\Delta \cup \Theta$ if $j(\Delta \cup \Theta) \approx \Delta \cup \Theta$ for any permutation j of the constants in $\mathcal{C}$.*

The set of maximal interchangeable subsets of a set of constants is the uniquely defined partition of this set and will be called the *interchangeable partition*. To check whether a set of constants $\mathcal{C}$ is interchangeable, it is sufficient to check that $j(\Delta \cup \Theta) \approx \Delta \cup \Theta$ for those permutations which swap just two constants from $\mathcal{C}$. Note that the constants do not actually need to appear in $\Delta \cup \Theta$. It trivially holds that constants which do not appear in $\Delta \cup \Theta$ are interchangeable. When $\mathcal{I} = \{\mathcal{C}_1, \ldots, \mathcal{C}_n\}$ is the interchangeable partition of a set of constants then we may introduce a new type $type_{lexmin(\mathcal{C}_i)}$ for every $\mathcal{C}_i \in \mathcal{I}$ (where $lexmin(\mathcal{C})$ denotes the lexically[1] smallest element from $\mathcal{C}$). When $D = \alpha \mid\!\sim \beta$ is a ground default rule, we write $variabilize(D)$ to denote the following default rule: $\bigwedge\{V_c \neq V_d : c, d \in const(D), \tau(c) = \tau(d)\} \wedge \alpha' \mid\!\sim \beta'$ where $const(D)$ is the set of constants appearing in D and α' and β' are obtained from α and β by respectively replacing all constants c by a new variable V_c of type $\tau(c)$. Here $\neq$ is treated as a binary predicate which is defined in the set of hard rules Θ.

Let $\mathcal{C}$ be a set of constants and let $\mathcal{I} = \{\mathcal{C}_1, \ldots, \mathcal{C}_n\}$ be the interchangeable partition of the constants from $\mathcal{C}$. Two ground default rules $\alpha_1 \mid\!\sim \beta_1$ and $\alpha_2 \mid\!\sim \beta_2$ are said to be *weakly isomorphic* w.r.t. $\mathcal{I}$ if $variabilize(\alpha_1 \mid\!\sim \beta_1)$ and $variabilize(\alpha_2 \mid\!\sim \beta_2)$ are isomorphic[2].

Definition 3 (Ground representatives). *Let $D = \alpha \mid\!\sim \beta$ be a default rule and let $\mathcal{I} = \{\mathcal{C}_1, \ldots, \mathcal{C}_n\}$ be the interchangeable partition of constants. A set of ground representatives of D w.r.t. $\mathcal{I}$ is a maximal set of groundings of D by constants from $\bigcup_{\mathcal{C}_i \in \mathcal{I}} \mathcal{C}_i$ such that no two of these groundings are weakly isomorphic w.r.t. $\mathcal{I}$. (If $\alpha \mid\!\sim \beta$ is typed then we only consider groundings which respect the typing of variables.)*

A set of ground representatives of a default rule $D = \alpha \mid\!\sim \beta$ can be constructed in time $O(|\mathcal{I}|^{|D|})$. While this is exponential in the size of the default rule (which is usually small), it is only polynomial in the number of classes in the interchangeable partition $\mathcal{I}$ and does not depend on the total number of constants.

---

[1] Here, we are just ordering the constants by the lexical ordering of their names.

[2] We will omit "w.r.t. $\mathcal{I}$" when it is clear from the context.

Let $\Delta \cup \Theta$ be a first-order default theory and $\mathcal{C}$ a set of constants. Let $\mathcal{R} = \bigcup_{\alpha \vdash\!\!\sim \beta \in \Delta} \mathcal{R}_{\alpha \vdash\!\!\sim \beta}$ where $\mathcal{R}_{\alpha \vdash\!\!\sim \beta}$ denotes a set of ground representatives of $\alpha \vdash\!\!\sim \beta$. The rational closure for the first-order default theory is based on the partition[3] $\Delta_1^* \cup \dots \cup \Delta_k^*$ of the set

$$\Delta^* = \{variabilize(\alpha \vdash\!\!\sim \beta) : \alpha \vdash\!\!\sim \beta \in \mathcal{R} \text{ and } \{\alpha\} \cup \Theta \not\vdash \bot\}$$

where Δ_j^* is the set of default rules $variabilize(\alpha \vdash\!\!\sim \beta) \in \Delta^* \setminus (\Delta_1^* \cup \dots \cup \Delta_{j-1}^*)$ such that

$$\{\alpha \wedge \beta\} \cup \{\neg\alpha_i \vee \beta_i : (\alpha_i \vdash\!\!\sim \beta_i) \in \Delta^* \setminus (\Delta_1^* \cup \dots \cup \Delta_{j-1}^*)\} \cup \Theta \qquad (3)$$

has a model with Herbrand universe $\mathcal{C}$. When all rules $\alpha' \vdash\!\!\sim \beta'$ from the set

$$\Delta_{\alpha \vdash\!\!\sim \beta}^* = \{variabilize(\alpha' \vdash\!\!\sim \beta') : \alpha' \vdash\!\!\sim \beta' \text{ is a ground representative of } \alpha \vdash\!\!\sim \beta\},$$

are contained in the same partition class Δ_j^* then we can simplify Δ_j^* by setting $\Delta_j^* := \Delta_j^* \cup \{\alpha \vdash\!\!\sim \beta\} \setminus \Delta_{\alpha \vdash\!\!\sim \beta}^*$. Furthermore, note that checking the existence of Herbrand models can be carried out using cutting-plane inference which means that it is seldom needed to ground the set of default rules completely. We can now present the lifted counterparts to Transformations 1–3.

Transformation 4 (Lifted rational closure). *Let Δ be a set of default rules and let Θ be a set of hard constraints. Let $\Delta_1^* \cup \dots \cup \Delta_k^*$ be the partition of $\Delta \cup \Theta$, defined by (3). Let the MLN $\mathcal{M}$ be defined as follows: $\bigcup_{i=1}^k \{(\neg a_i \vee \neg\alpha \vee \beta, \infty) : \alpha \vdash\!\!\sim \beta \in \Delta_i^*\} \cup \{(a_i, 1)\} \cup \{(\phi, \infty) : \phi \in \Theta\} \cup \{(a_i \vee \neg a_{i-1}, \infty)\}$ where a_i are auxiliary (ground) literals. If (Δ, Θ) is satisfiable then $\mathcal{M}$ is a Markov logic model of (Δ, Θ).*

Transformation 5 (Lifted lexicographic entailment). *Let Δ be a set of default rules, let Θ be a set of hard constraints, and let $\mathcal{U}$ be the considered set of constants. Let $\Delta_1^* \cup \dots \cup \Delta_k^*$ be the partition of $\Delta \cup \Theta$, defined by (3). Let the MLN $\mathcal{M}$ be defined as follows: $\bigcup_{i=1}^k \{(\neg\alpha \vee \beta, \lambda_i) : \alpha \vdash\!\!\sim \beta \in \Delta_i\} \cup \{(\phi, \infty) : \phi \in \Theta\}$ where $\lambda_i = 1 + \sum_{j=1}^{i-1} \sum_{\alpha \vdash\!\!\sim \beta \in \Delta_j^*} |\mathcal{U}|^{|vars(\alpha \vdash\!\!\sim \beta)|} \cdot \lambda_j$ for $i > 1$ and $\lambda_1 = 1$. If (Δ, Θ) is satisfiable then $\mathcal{M}$ is a Markov logic model of (Δ, Θ).*

Note that lexicographic entailment may lead to MLNs with very large weights.[4]

Next, we describe a lifted variant of maximum-entropy entailment. Let $\Delta \cup \Theta$ be a first-order default theory and $\mathcal{I}$ the interchangeable partition of constants from a given set $\mathcal{C}$. Let $\Delta_1^* \cup \dots \cup \Delta_k^*$ be the partition of $\Delta \cup \Theta$, defined as in (3) (without the simplification of merging default rules), and let $\Gamma := \Delta_1^*$. First, we construct an MLN $\mathcal{M}$ containing the rules from Γ and set their weights

---

[3] With a slight abuse of terminology, we will call $\Delta_1^* \cup \dots \cup \Delta_k^*$ the *partition of $\Delta \cup \Theta$* even though it is strictly speaking only a partition of Δ^*.

[4] Although existing MLN systems are not able to work with weights as large as are sometimes produced, due to numerical issues, we have implemented an MLN system based on cutting-plane MAP inference which can work with arbitrarily large weights.

equal to 1. For every Δ_j^* with $j > 1$, while $\Delta_j^* \not\subseteq \Gamma$, we repeat the following steps. We construct a new MLN $\mathcal{M}'$ by adding to the MLN $\mathcal{M}$ all rules from the set $\{\neg\alpha \vee \beta \colon \alpha \mathrel|\!\sim \beta \in (\Delta_j^* \cup \ldots \Delta_k^*) \setminus \Gamma\}$ as hard constraints (i.e. with infinite weights). For every $\alpha \mathrel|\!\sim \beta \in \Delta_j^* \setminus \Gamma$, we construct its ground representative $\alpha' \mathrel|\!\sim \beta'$ (note that there is only one ground representative up to isomorphism for any rule in Δ_j^*, which follows from the construction of Δ_j^*) and we find a most probable world $\omega_{\alpha \mathrel|\!\sim \beta}$ of $(\mathcal{M}', \alpha')$; let us write $p_{\alpha \mathrel|\!\sim \beta}$ for its penalty, i.e. the sum of the weights of the violated rules. Note that $\omega_{\alpha \mathrel|\!\sim \beta}$ verifies the default rule $\alpha \mathrel|\!\sim \beta$ and only falsifies rules in Γ, exactly as in the propositional version of the maximum-entropy transformation. We then select the rules $\alpha \mathrel|\!\sim \beta$ with minimum penalty $p_{\alpha \mathrel|\!\sim \beta}$, add them to Γ and to the MLN $\mathcal{M}$ with the weight set to $1 + p_{\alpha \mathrel|\!\sim \beta}$. If $\mathcal{M}'$ does not have any models, the initial set of defaults cannot be satisfiable, and we end the procedure.

Transformation 6 (Lifted maximum-entropy entailment). *Let Δ be a set of default rules, let Θ be a set of hard constraints, and let $\mathcal{U}$ be the considered set of constants. Let $\mathcal{M}$ be the MLN obtained in the last iteration of the procedure described above. If (Δ, Θ) is satisfiable then $\mathcal{M}$ is a Markov logic model of (Δ, Θ).*

Example 2. Let us consider the following defaults:

$$bird(X) \mathrel|\!\sim flies(X) \qquad bird(X) \wedge antarctic(X) \mathrel|\!\sim \neg flies(X)$$
$$bird(X) \wedge antarctic(X) \wedge (X \neq Y) \wedge sameSpecies(X, Y) \mathrel|\!\sim antarctic(Y)$$

Let the set of constants be given by $\mathcal{C} = \{tweety, donald, beeper\}$. Then the lexicographic transformation yields the MLN $\{(\phi_1, 1), (\phi_2, 4), (\phi_3, 4)\}$ while the maximum entropy transformation yields $\{(\phi_1, 1), (\phi_2, 2), (\phi_3, 3)\}$, where $\phi_1 = \neg bird(X) \vee flies(Y)$, $\phi_2 = \neg bird(X) \vee \neg sameSpecies(X, Y) \vee \neg(X \neq Y) \vee \neg bird(Y) \vee \neg antarctic(X) \vee antarctic(Y)$ and $\phi_3 = \neg bird(X) \vee \neg antarctic(X) \vee \neg flies(X)$.

As the next example illustrates, it is sometimes necessary to split the initial default rules into several typed specializations.

Example 3. Consider the following defaults: $bird(X) \wedge (X \neq tweety) \mathrel|\!\sim flies(X)$, $bird(X) \wedge antarctic(X) \mathrel|\!\sim \neg flies(X)$ and $bird(X) \wedge antarctic(X) \wedge (X \neq Y) \wedge sameSpecies(X, Y) \mathrel|\!\sim antarctic(Y)$. Then the lexicographic transformation yields the MLN $\{(\phi_1, 1), (\phi_2, 1), (\phi_3, 7), (\phi_4, 7)\}$, where:

$$\phi_1 = \neg bird(\tau_{tweety} : X) \vee \neg antarctic(\tau_{tweety} : X) \vee \neg flies(\tau_{tweety} : X),$$
$$\phi_2 = \neg bird(\tau_{beeper} : X) \vee \neg(\tau_{beeper} : X \neq \tau_{tweety} : tweety) \vee flies(\tau_{beeper} : X),$$
$$\phi_3 = \neg bird(\tau_{beeper} : X) \vee \neg antarctic(\tau_{beeper} : X) \vee \neg flies(\tau_{beeper} : X),$$
$$\phi_4 = \neg bird(X) \vee \neg sameSpecies(X, Y) \vee \neg(X \neq Y), \neg bird(Y) \vee \neg antarctic(X)$$

Note that the transformation had to introduce new types corresponding to the interchangeable sets of constants $\{\{tweety\}, \{beeper, donald\}\}$. The rule ϕ_4 was created by merging rules with different typing, which was made possible by the fact that all the respective differently typed rules ended up with the same weights. The maximum entropy transformation leads to six such rules.

5   Evaluation

In this section we describe experimental evaluation of the methods presented in this paper. The implementation of the described methods is available for download[5].

We have evaluated the proposed methods using the well-known UW-CSE dataset, which describes the Department of Computer Science and Engineering at the University of Washington [20]. The usual task is to predict the *advisedBy(person, person)* predicate from the other predicates. A set of rules for this domain has previously been collected for the experiments in [20]. These rules, however, cannot be used as default rules because they are not satisfiable in the sense of Definition 1. Therefore, in order to evaluate our method, we have used the following consistent set of default rules.

D_1 : $\mathrel{|\!\sim} \neg advisedBy(S, P)$

D_2 : $advisedBy(S, P_1) \mathrel{|\!\sim} \neg tempAdvisedBy(S, P_2)$

D_3 : $advisedBy(S, P) \wedge publication(Pub, S) \mathrel{|\!\sim} publication(Pub, P)$

D_4 : $(P_1 \neq P_2) \wedge advisedBy(S, P_1) \mathrel{|\!\sim} \neg advisedBy(S, P_2)$

D_5 : $advisedBy(S, P) \wedge ta(C, S, T) \mathrel{|\!\sim} taughtBy(C, P, T)$

D_6 : $professor(P) \wedge student(S) \wedge publication(Pub, P) \wedge publication(Pub, S)$
$\mathrel{|\!\sim} advisedBy(S, P)$

D_7 : $professor(P) \wedge student(S) \wedge publication(Pub, P) \wedge publication(Pub, S) \wedge$
$tempAdvisedBy(S, P2) \mathrel{|\!\sim} \neg advisedBy(S, P)$

D_8 : $(S_1 \neq S_2) \wedge advisedBy(S_2, P) \wedge ta(C, S_2, T) \wedge ta(C, S_1, T) \wedge$
$taughtBy(C, P, T) \wedge student(S_1) \wedge professor(P) \mathrel{|\!\sim} advisedBy(S_1, P)$

D_9 : $(S_1 \neq S_2) \wedge advisedBy(S_2, P) \wedge ta(C, S_2, T) \wedge ta(C, S_1, T) \wedge$
$taughtBy(C, P, T) \wedge student(S_1) \wedge professor(P) \wedge tempAdvisedBy(S_1, P_2)$
$\mathrel{|\!\sim} \neg advisedBy(S_1, P)$

Recall that default rules $\alpha \mathrel{|\!\sim} \beta$ in our setting correspond to statements of the form: *for any grounding substitution θ, $\beta\theta$ is true in all most probable worlds of* $(\mathcal{M}, \alpha\theta)$. Thus the default rules $\alpha \mathrel{|\!\sim} \beta$ we consider should be such that an expert believes that $\alpha\theta$ being the only evidence, it would make sense to conclude $\beta\theta$. Seen with this perspective in mind, rule D_1 states that in absence of any knowledge, we assume that persons S and P are not in the *advisedBy* relationship. Rule D_2 states that if we only know that S has an advisor then we conclude that S does not have a temporary advisor. Rule D_3 states that advisors are typically co-authors of their students' papers. Rule D_4 states that students typically only have one advisor. The rest of the rules can be interpreted similarly. Note that rules D_7 and D_9 encode exceptions to rules D_6 and D_8.

---

[5] https://github.com/supertweety/mln2poss.

We computed the lexicographic and maximum-entropy transformations of these rules using our implementation of the described methods.[6] We evaluated the obtained MLNs on the five different subject areas of the UW-CSE dataset, which is the standard methodology. Specifically, we computed the average number of true positives and false positives[7] for the *advisedBy* predicate over 10 runs of MAP inference, noting that the results can depend on the specific MAP state that is returned. For comparison, we have used an MLN with the same set[8] of rules but with weights learned discriminatively using Tuffy [15] (LEARNED), and an MLN with the same set of rules but with all weights set to 1 (ONES). The results are shown in Table 1. The maximum entropy and lexicographic entailment have highest recall but at the cost of also having a higher number of false positives. Note that the number of pairs which can potentially be in the *advisedBy* relationship is in the order of hundreds or even thousands but the true number of pairs of people in this relationship is in the order of just tens. The baseline method ONES has largest variance.

Table 1. Experimental results for MLNs obtained by the described methods (the numbers represent absolute counts).

	MaxEnt		LEX		ONES		LEARNED	
	TP	FP	TP	FP	TP	FP	TP	FP
AI	10 ± 0	7 ± 0	10 ± 0	7 ± 0	8.6 ± 0.7	4.9 ± 0.9	10 ± 0	2 ± 0
GRA	4 ± 0	5 ± 0	4 ± 0	5 ± 0	3.5 ± 0.7	3.9 ± 0.7	2 ± 0	2 ± 0
LAN	0 ± 0	0 ± 0	0 ± 0	0 ± 0	0 ± 0	0 ± 0	2 ± 0	1 ± 0
SYS	10.5 ± 0.5	3.5 ± 0.5	11 ± 0	3 ± 0	7.2 ± 1.1	2.4 ± 0.5	4 ± 0	0 ± 0
THE	3 ± 0	3 ± 0	3 ± 0	3 ± 0	3 ± 0	1.7 ± 0.7	2 ± 0	1 ± 0

6   Conclusion

We have discussed the problem of constructing a Markov logic network (MLN) from a set of first-order default rules, where default rules are seen as constraints

---

[6] Our implementation is based on a cutting-plane inference method for MAP inference implemented using the SAT4J library [5] and the MLN system Tuffy [15].

[7] Note that using AUC as an evaluation metric would not make sense in this case because of the way the MLNs are constructed by our approach. The construction can produce MLNs which make sensible predictions when used together with MAP inference but which do not have to be meaningful for the given datasets as probability distributions. After all, our MLN construction methods do not assume any information from which the probabilities could be inferred, except qualitative information on rankings of possible worlds expressed by default rules.

[8] We had to remove D_3 for efficiency reasons, though.

on what should be derivable using MAP inference. The proposed construction methods have been obtained by lifting three well-known methods for non-monotonic reasoning about propositional default rules: the rational closure, the lexicographic closure and the maximum-entropy closure. As our evaluation with the UW-CSE dataset illustrates, our method can be used to construct useful MLNs in scenarios where no training data is available. In the future, we would like to explore the connections between our proposed lifted transformations and the lifted inference literature. For example, identifying interchangeable constants is known as shattering in lifted inference [6].

Acknowledgement. We thank the anonymous reviewers for their detailed comments. This work has been supported by a grant from the Leverhulme Trust (RPG-2014-164).

A   Proofs

Here we provide formal justifications for the transformations presented in this paper[9]. We start by proving correctness of the ground transformations.

Proposition 1. *Let Δ be a set of default rules. Let Δ^{rat}, Δ^{lex} and Δ^{ent} be the rational, lexicographic and maximum entropy closure of Δ, respectively. Let $\mathcal{M}^{rat}$, $\mathcal{M}^{lex}$ and $\mathcal{M}^{ent}$ be Markov logic networks obtained from Δ by Transformations 1, 2 and 3, respectively. Then the following holds for any default rule $\alpha \mathrel{|\!\sim} \beta$:*

1. *$\alpha \mathrel{|\!\sim} \beta \in \Delta^{rat}$ if and only if $(\mathcal{M}^{rat}, \{\alpha\}) \vdash_{MAP} \beta$,*
2. *$\alpha \mathrel{|\!\sim} \beta \in \Delta^{lex}$ if and only if $(\mathcal{M}^{lex}, \{\alpha\}) \vdash_{MAP} \beta$,*
3. *$\alpha \mathrel{|\!\sim} \beta \in \Delta^{ent}$ if and only if $(\mathcal{M}^{ent}, \{\alpha\}) \vdash_{MAP} \beta$.*

Proof. Throughout the proof, let $\Delta = \Delta_1 \cup \Delta_2 \cup \cdots \cup \Delta_k$ be the Z-ordering of Δ.

1. Let $\alpha \mathrel{|\!\sim} \beta \in \Delta^{rat}$ be a default rule. Let j be the smallest index such that $\Delta^{rat}_\alpha = \{\neg \gamma \vee \delta | \gamma \mathrel{|\!\sim} \delta \in \Delta_j \cup ... \cup \Delta_k\} \cup \{\alpha\}$ is consistent. Recall that $\alpha \mathrel{|\!\sim} \beta \in \Delta^{rat}$ if and only if $\Delta^{rat}_\alpha \models \beta$. By the construction of the MLN $\mathcal{M}^{rat}$ it must hold that $(\mathcal{M}^{rat}, \{\alpha\}) \vdash_{MAP} \neg a_i$ for every $i < j$ and also $(\mathcal{M}^{rat}, \{\alpha\}) \vdash_{MAP} a_i$ for all $i \geq j$. Therefore all $\neg \alpha \vee \beta$, such that $\alpha \mathrel{|\!\sim} \beta \in \Delta_i$ where $i \geq j$, must be true in all most probable worlds of $(\mathcal{M}^{rat}, \{\alpha\})$. But then necessarily we have: if $\Delta^{rat}_\alpha \models \beta$ then $(\mathcal{M}^{rat}, \{\alpha\}) \vdash_{MAP} \beta$. Similarly, to show the other direction of the implication, let us assume that $(\mathcal{M}^{rat}, \{\alpha\}) \vdash_{MAP} \beta$. Then we can show using basically the identical reasoning as for the other direction that the set of formulas $\neg \alpha \vee \beta$ which must be satisfied in all most probable worlds of $(\mathcal{M}^{rat}, \{\alpha\})$ is equivalent to the set of formulas in Δ^{rat}_α.

   2. It holds that $\alpha \mathrel{|\!\sim} \beta \in \Delta^{lex}$ if and only if β is true in all lex-preferred models of α, i.e. $\forall \omega \in [\![\alpha]\!] : (\omega \not\models \beta) \Rightarrow \exists \omega' \in [\![\alpha]\!] : \omega' \prec \omega$ where $\prec$ is the lex-preference relation based on Z-ordering defined in Sect. 2.2. What we need to show is that for

---

[9] For brevity we omit hard rules here because generalizations of the proofs to involve hard rules are rather straightforward, but a bit too verbose.

any possible worlds ω, ω' it holds $\omega \prec \omega'$ if and only $P_{\mathcal{M}^{lex}}(\omega) > P_{\mathcal{M}^{lex}}(\omega')$ where $P_{\mathcal{M}^{lex}}$ is the probability given by the MLN $\mathcal{M}^{lex}$, from which correctness of the lexicographic transformation will follow. But this actually follows immediately from the way we set the weights in Transformation 2, as the *penalty* for not satisfying one formula corresponding to a default rule in Δ_i is greater than the sum of *penalties* for not satisfying all formulas corresponding to the default rules in $\bigcup_{j<i} \Delta_j$.

3. This follows directly from the results in the paper [10] in which maximum entropy closure was introduced. An explicit ranking function on possible worlds was derived in that paper, which we explicitly use in the transformation.

Next we show correctness of the non-ground transformations. We start by proving properties of the non-ground counterpart of Z-ordering.

Proposition 2. *Let Δ^* be a default theory and $\mathcal{C}$ be a set of constants (universe). Let Δ be the set of groundings[10] of default rules from Δ^*. Let $\Delta = \Delta_1 \cup \cdots \cup \Delta_k$ be Z-ordering of the set of ground default rules Δ. Let $\Delta_1^* \cup \cdots \cup \Delta_k^*$ be as defined by Eq. 3. Then a ground default rule $\alpha \mathbin{|\!\sim} \beta$ is in Δ_i if and only if a rule isomorphic to variabilize($\alpha \mathbin{|\!\sim} \beta$) is in Δ_i^*.*

Proof. (Sketch) This proposition follows from the simple observation that Eq. 3 is equivalent to checking whether the ground default rule $\alpha \mathbin{|\!\sim} \beta$ is tolerated by the set of groundings of the default rules $\gamma \mathbin{|\!\sim} \delta \in \Delta^* \setminus (\Delta_1^* \cup \cdots \cup \Delta_{j-1}^*)$ (because we explicitly ask there about existence of a Herbrand model[11] with universe $\mathcal{C}$). Since the answer, whether it is tolerated or not, must be the same for every default rule weakly isomorphic to $\alpha \mathbin{|\!\sim} \beta$, it follows that this is equivalent to checking this condition for all groundings of variabilize($\alpha \mathbin{|\!\sim} \beta$), which must then necessarily give us an equivalent result to what we would obtain by Z-ordering performed on the explicitly enumerated groundings. The statement of the proposition then follows from this.

In other words, what the above proposition states, is that if we replace non-ground rules in the particular Δ_i^*'s by all their groundings then this partitioning of ground default rules must be equivalent to what we would obtain by directly Z-ordering the ground default rules in the set $\mathcal{R}$.

Proposition 3. *Let Δ^* be a set of non-ground default rules and $\mathcal{C}$ be a set of constants (universe). Let Δ^{rat}, Δ^{lex} and Δ^{ent} be the rational, lexicographic and maximum entropy closure, respectively, of the set of default rules obtained by grounding Δ^*. Let $\mathcal{M}^{rat}$, $\mathcal{M}^{lex}$ and $\mathcal{M}^{ent}$ be Markov logic networks obtained from Δ^* by Transformations 4, 5 and 6, respectively. Then the following holds for any ground default rule $\alpha \mathbin{|\!\sim} \beta$:*

---

[10] Recall that the formulas in Δ^* are typed according to interchangeability of constants. The groundings must respect the typing information. This will be the case whenever we speak of groundings in this section.

[11] However, this does not mean that we need to ground this theory completely in order to solve it, e.g. by using cutting plane inference we can avoid the need to ground it completely.

1. $\alpha \mid\sim \beta \in \Delta^{rat}$ *if and only if* $(\mathcal{M}^{rat}, \{\alpha\}) \vdash_{MAP} \beta$,
2. $\alpha \mid\sim \beta \in \Delta^{lex}$ *if and only if* $(\mathcal{M}^{lex}, \{\alpha\}) \vdash_{MAP} \beta$,
3. $\alpha \mid\sim \beta \in \Delta^{ent}$ *if and only if* $(\mathcal{M}^{ent}, \{\alpha\}) \vdash_{MAP} \beta$.

Proof. 1. This case follows from Propositions 1 and 2 by noticing that the constructed MLNs, when grounded, are the same as what the ground Transformation 1 would produce if applied on all groundings of default rules from Δ^*.

2. From Proposition 2 we have that, if we ground the MLN produced by Transformation 5, then the structure of the MLN will be identical to what we would obtain if we applied Transformation 2 on all groundings of default rules from Δ^*. While the weights of the formulas are not the same, it is still guaranteed that $\omega \prec \omega'$ if and only $P_{\mathcal{M}^{lex}}(\omega) > P_{\mathcal{M}^{lex}}(\omega')$, where $\prec$ is the lex-preference relation. This is because the term $|\mathcal{C}|^{|vars(\alpha \mid\sim \beta)|}$, which is used to define the weights in Transformation 5, is an upper bound on the number of groundings of a default rule $\alpha \mid\sim \beta$ (this implies that the sum of all weights of groundings of formulas in the MLN which correspond to default rules from $\bigcup_{i<j} \Delta_i$ will be smaller than the weight of a single formula corresponding to a default rule from Δ_j which is what we need).

3. (Sketch) To show the last part of this proposition, we would basically need to replicate a more detailed reasoning from the proof of Proposition 2 because maximum entropy closure needs to create a partitioning of the set of default rules which refines Z-ordering. Since no new ideas are needed for this proof and because of space limitations, we omit details. The basic idea is the same as for the non-ground Z-ordering – we only process representatives of the non-ground default rules and we can show that we would obtain an equivalent result if we processed all groundings of the default rules by the procedure from Transformation 3.

References

1. Benferhat, S., Bonnefon, J.F., da Silva Neves, R.: An overview of possibilistic handling of default reasoning, with experimental studies. Synthese **146**(1–2), 53–70 (2005)
2. Benferhat, S., Dubois, D., Prade, H.: Nonmonotonic reasoning, conditional objects and possibility theory. Artif. Intell. **92**(1–2), 259–276 (1997)
3. Benferhat, S., Dubois, D., Prade, H.: Practical handling of exception-tainted rules and independence information in possibilistic logic. Appl. Intell. **9**(2), 101–127 (1998)
4. Benferhat, S., Dubois, D., Prade, H.: Possibilistic and standard probabilistic semantics of conditional knowledge bases. J. Log. Comput. **9**(6), 873–895 (1999)
5. Berre, D.L., Parrain, A.: The Sat4j library, release 2.2. J. Satisfiability, Boolean Model. Comput. **7**, 50–64 (2010)
6. de Salvo Braz, R., Amir, E., Roth, D.: Lifted first-order probabilistic inference. In: Kaelbling, L.P., Saffiotti, A. (eds.) Proceeding of the 19th Joint Conference on Artificial Intelligence, p. 1319 (2005)
7. de Saint-Cyr, F.D., Lang, J., Schiex, T.: Penalty logic and its link with Dempster-Shafer theory. In: Proceedings of the 10th International Conference on Uncertainty in Artificial Intelligence, pp. 204–211 (1994)

8. Friedman, N., Halpern, J.Y., Koller, D.: First-order conditional logic for default reasoning revisited. ACM Trans. Comput. Log. **1**(2), 175–207 (2000)
9. Geffner, H., Pearl, J.: Conditional entailment: bridging two approaches to default reasoning. Artif. Intell. **53**(2), 209–244 (1992)
10. Goldszmidt, M., Morris, P., Pearl, J.: A maximum entropy approach to nonmonotonic reasoning. IEEE Trans. Pattern Anal. Mach. Intell. **15**(3), 220–232 (1993)
11. Halpern, J.Y.: An analysis of first-order logics of probability. Artif. Intell. **46**(3), 311–350 (1990)
12. Kern-Isberner, G., Thimm, M.: A ranking semantics for first-order conditionals. In: Proceedings of the 20th European Conference on Artificial Intelligence, pp. 456–461 (2012)
13. Kraus, S., Lehmann, D., Magidor, M.: Nonmonotonic reasoning, preferential models and cumulative logics. Artif. Intell. **44**(1–2), 167–207 (1990)
14. Lehmann, D., Magidor, M.: What does a conditional knowledge base entail? Artif. Intell. **55**(1), 1–60 (1992)
15. Niu, F., Ré, C., Doan, A., Shavlik, J.W.: Tuffy: scaling up statistical inference in markov logic networks using an RDBMS. PVLDB **4**(6), 373–384 (2011)
16. Noessner, J., Niepert, M., Stuckenschmidt, H., Rockit.: Exploiting parallelism and symmetry for map inference in statistical relational models. In: Proceedings of the 27th Conference on Artificial Intelligence, AAAI (2013)
17. Pápai, T., Ghosh, S., Kautz, H.: Combining subjective probabilities and data in training markov logic networks. In: Flach, P.A., De Bie, T., Cristianini, N. (eds.) ECML PKDD 2012, Part I. LNCS, vol. 7523, pp. 90–105. Springer, Heidelberg (2012)
18. Pearl, J.: Probabilistic reasoning in intelligent systems: networks of plausible inference. Morgan Kaufmann, Burlington (1988)
19. Pearl, J., System, Z.: A natural ordering of defaults with tractable applications to nonmonotonic reasoning. In: Proceedings of the 3rd Conference on Theoretical Aspects of Reasoning about Knowledge, pp. 121–135 (1990)
20. Richardson, M., Domingos, P.: Markov logic networks. Mach. Learn. **62**(1–2), 107–136 (2006)
21. Riedel, S.: Improving the accuracy and efficiency of MAP inference for Markov logic. In: Proceedings of the Twenty-Fourth Conference on Uncertainty in Artificial Intelligence, pp. 468–475 (2008)
22. Tversky, A., Kahneman, D.: Extensional versus intuitive reasoning: the conjunction fallacy in probability judgment. Psychol. Rev. **90**(4), 293 (1983)

Mine 'Em All: A Note on Mining All Graphs

Ondřej Kuželka[1]([✉]) and Jan Ramon[2,3]

[1] School of Computer Science & Informatics, Cardiff University, Cardiff, UK
KuzelkaO@cardiff.ac.uk
[2] Department of Computer Science, KU Leuven, Leuven, Belgium
jan.ramon@cs.kuleuven.be
[3] INRIA, Lille, France

Abstract. We study the complexity of the problem of enumerating all graphs with frequency at least 1 and computing their support. We show that there are hereditary classes of graphs for which the complexity of this problem depends on the order in which the graphs should be enumerated (e.g. from frequent to infrequent or from small to large). For instance, the problem can be solved with polynomial delay for databases of planar graphs when the enumerated graphs should be output from large to small but it cannot be solved even in incremental-polynomial time when the enumerated graphs should be output from most frequent to least frequent (unless P=NP).

1 Introduction

In this paper we study graph mining problems from a nontraditional perspective. We are inspired by the question which properties of the problem make some graph mining problems solvable in incremental polynomial time or with polynomial delay. Here, we do not require the discovered graph patterns to be frequent and we want to output all patterns occurring in at least one database graph. However, we still want to also output their occurrences. In addition, we constrain the order in which the patterns should be printed, e.g. from most frequent patterns to least frequent patterns, which allows us to connect our results to results on (in)frequent graph mining. Surprisingly, for several graph classes, we show that different orders lead to very different computational complexities. For instance mining planar graphs cannot be done *in incremental-polynomial time* when the output graphs should be ordered by frequency but it can be done with *polynomial delay* when they should be ordered from largest to smallest.

2 Preliminaries

In this section we first briefly review some basic concepts and fix the notations used in this paper. We start with some standard definitions from graph theory.

© Springer International Publishing Switzerland 2016
K. Inoue et al. (Eds.): ILP 2015, LNAI 9575, pp. 106–121, 2016.
DOI: 10.1007/978-3-319-40566-7_8

Graphs. An *undirected graph* is a pair (V, E), where V is a finite set of *vertices* and $E \subseteq \{e \subseteq V : |e| = 2\}$ is a set of *edges*. Two vertices are said to be *adjacent* (in G) if they are connected by an edge (of the graph G). A *labeled undirected graph* is a triple (V, E, λ), where (V, E) is an undirected graph and $\lambda : V \cup E \to \Sigma$ is a function assigning a label from an alphabet Σ to every element of $V \cup E$. We will denote the set of vertices, the set of edges, and the labeling function of a graph G by $V(G)$, $E(G)$, and λ_G, respectively. We define $|G| = |V(G)| + |E(G)|$ and call this the order of G. Note that in graph theory, often other notions of 'order' are used, measuring only the number of edges or the number of vertices. A graph G' is a *subgraph* of a graph G, if $V(G') \subseteq V(G)$, $E(G') \subseteq E(G)$, and $\lambda_{G'}(x) = \lambda_G(x)$ for every $x \in V(G') \cup E(G')$; G' is an *induced subgraph* of G if it is a subgraph of G satisfying $\{u, v\} \in E(G')$ if and only if $\{u, v\} \in E(G)$ for every $u, v \in V(G')$. For a subset $S \subseteq V(G)$, $G[S]$ denotes the (unique) induced subgraph of G with vertex set S. A *contraction* of an edge $e = \{u, v\}$ in a graph G is an operation which produces a new graph by replacing u and v in $V(G)$ as well as in all $\{x, y\} \in E(G)$ by a new vertex w (pictorially, this can be imagined as shrinking the edge). A *subdivision* of an edge $e = \{u, v\}$ is an operation which produces a new graph by removing e and adding a path connecting u and v.

Tree Decomposition, Tree-width. The notion of tree-width was reintroduced in [3,11]. It proved to be a useful parameter of graphs in algorithmic graph theory. A *tree-decomposition* of a graph G, denoted $TD(G)$, is a pair $(T, \mathcal{X})$, where T is a rooted unordered tree and $\mathcal{X} = (X_z)_{z \in V(T)}$ is a family of subsets of $V(G)$ satisfying

(i) $\cup_{z \in V(T)} X_z = V(G)$,
(ii) for every $\{u, v\} \in E(G)$, there is a $z \in V(T)$ such that $u, v \in X_z$, and
(iii) $X_{z_1} \cap X_{z_3} \subseteq X_{z_2}$ for every $z_1, z_2, z_3 \in V(T)$ such that z_2 is on the simple path connecting z_1 with z_3 in T.

The set X_z associated with a node z of T is called the *bag* of z. The nodes of T will often be referred to as the nodes of $TD(G)$. The tree-width of $TD(G)$ is $\max_{z \in V(T)} |X_z| - 1$, and the *tree-width* of G, denoted $\mathrm{tw}(G)$, is the minimum tree-width over all tree-decompositions of G. By graphs of bounded tree-width' we mean graphs of tree-width at most k, where k is some constant.

A class of graphs $\mathcal{G}$ is called *hereditary* if for any graph $G \in \mathcal{G}$ all its subgraphs also belong to $\mathcal{G}$. The class of graphs of treewidth at most k is hereditary. The same also holds for planar graphs (as clearly any subgraph of a planar graph is still planar).

Graph Isomorphism, Graph Canonization. Graphs G and G' are *isomorphic* if there exists a bijection $\pi : V(G) \to V(G')$ such that $\{u, v\} \in E(G)$ if and only if $\{\pi(u), \pi(v)\} \in E(G')$. Graph canonization is a function from graphs to strings such that two graphs have the same canonization if and only if they are isomorphic.

Subgraph Isomorphism and Induced Subgraph Isomorphism. We say that a graph G_1 is *subgraph isomorphic* to a graph G_2 if G_1 is isomorphic to a subgraph of G_2. We say that a graph G_1 is *induced subgraph isomorphic* to a graph G_2 if G_1 is isomorphic to an induced subgraph of G_2. Deciding whether a graph is (induced) subgraph isomorphic to another graph is NP-complete and it remains NP-complete even for bounded-treewidth graphs [10]. Note that there are graph classes, e.g. bounded-treewidth graphs or planar graphs, for which isomorphism can be decided in polynomial time but for which subgraph isomorphism is NP-complete. We are mostly interested in such classes because for them it is not obvious whether *fast* graph mining algorithms exist.

Homeomorphism and Induced Homeomorphism. We say that a graph G_1 is *homeomorphic* to a graph G_2 if there is a graph G_1' which can be obtained from G_1 by subdividing its edges and G_1' is subgraph isomorphic to G_2. We say that a graph G_1 is *induced homeomorphic* to a graph G_2 if there is a graph G_1' which can be obtained from G_1 by subdividing its edges and G_1' is induced subgraph isomorphic to G_2. Deciding if a graph is homeomorphic to another graph is NP-complete even for graphs of bounded treewidth and unbounded maximum degree [10].

Minor Embedding and Induced Minor Embedding. We say that a graph G_1 is *minor-embeddable* to a graph G_2 if there is a graph G_1' isomorphic to G_1 which can be obtained from a subgraph of G_2 by contracting edges and deleting loops and multiple-edges thus produced. We say that a graph G_1 is *induced minor-embeddable* to a graph G_2 if there is a graph G_1' isomorphic to G_1 which can be obtained from an induced subgraph of G_2 by contracting edges and deleting loops and multiple-edges thus produced. Deciding if a graph is minor-embeddable to another graph is NP-complete even for graphs of bounded treewidth and unbounded maximum degree [10].

Fixed-Parameter Tractability. Formally, a parameterized decision problem is a language $L \subseteq \Sigma^* \times N$ where Σ is a finite alphabet and N is the set of natural numbers [1]. An instance of a parametrized problem is a pair (x, k) where $k \in N$ is called *parameter* of the problem. A problem is *fixed-parameter tractable* (abbreviated *FPT*) if there exists an algorithm for solving instances of it which runs in time $|x|^{O(1)} \cdot f(k)$ where f is a computable function. Notice that whether a problem is fixed-parameter tractable depends on the selected parameterization. For instance, when the parameter of the problem is $|x|$, i.e. the actual size of the problem, then any problem e.g. from classes such as e.g. NP, EXP, NEXP is fixed-parameter tractable with such a parameterization. On the other hand, it is widely believed that e.g. the clique problem is not fixed-parameter tractable with the parameter being size of the clique. To capture a conjectured intractability hierarchy, the W-hierarchy is used which consists of an infinite number of increasingly more intractable classes W[1] , W[2], etc. The W-hierarchy is based on *fixed-parameter reductions*. A problem L_A is fixed-parameter reducible to a problem L_B if there exists an algorithm for transforming instances $(x,k)_A$ of the problem L_A to instances $(x',k')_B$ of the problem B such that:

(i) the transformation algorithm runs in time $|x|^{O(1)} \cdot f(k)$ where f is a computable function,
(ii) $k' \leq g(k)$ where g is a computable function,
(iii) $(x,k)_A \in L_A$ if and only if $(x',k')_B \in L_B$ (informally, $(x,k)_A$ has a 'yes' solution if and only if $(x',k')_B$ has a 'yes' solution).

3 Graph Mining Problems

In this section, we define the mining problems studied in this paper and describe their basic properties. We start with the definition of the classical frequent connected graph mining problem.

A *transaction database* is a a multiset of graphs from a given class $\mathcal{G}$. Given a pattern matching operator $\preccurlyeq$ (subgraph isomorphism or induced subgraph isomorphism), the frequency of a graph G in a transaction database DB, denoted by $\mathsf{freq}(G, DB)$, is given as $\mathsf{freq}(G, DB) = |\{G' \in DB | G \preccurlyeq G'\}|$. Given a threshold t, G is said to be frequent if $\mathsf{freq}(G, DB) \geq t$. The elements of the multiset $\{G' \in DB | G \preccurlyeq G'\}$ are called *occurrences* of the graph G in the database DB. We will often represent the set of occurrences also just by names or IDs of the graphs contained in it (e.g. G_1 will be represented just by "1").

Definition 1 (THE FREQUENT CONNECTED GRAPH MINING (FCGM) PROBLEM). *Given a class $\mathcal{G}$ of graphs, a transaction database DB of connected graphs from $\mathcal{G}$, a pattern matching operator $\preccurlyeq$, and frequency threshold, list the set of frequent connected graphs $G \in \mathcal{G}$ and their occurrences.*

In this paper, we are interested in another closely related type of problem which is to mine all graphs with frequency at least one *in certain order*.

Definition 2 (THE ORDERED MINING PROBLEMS). *Given a class $\mathcal{G}$ of graphs, a transaction database DB of connected graphs from $\mathcal{G}$ and a pattern matching operator $\preccurlyeq$, list the set of connected graphs $G \in \mathcal{G}$ with $\mathsf{freq}(G, DB) \geq 1$ and their occurrences in the transactions in the given order[1]:*

- *from most frequent to least frequent (ALL$_{F \to I}$ problem),*
- *from least frequent to most frequent (ALL$_{I \to F}$ problem),*
- *from smallest size to largest size (ALL$_{S \to L}$ problem),*
- *from largest size to smallest size (ALL$_{L \to S}$ problem).*

Here size of a graph G refers to $|E(G)|$ when $\preccurlyeq$ is subgraph isomorphism and to $|V(G)|$ when $\preccurlyeq$ is induced subgraph isomorphism.

---

[1] ALL$_{F \to I}$ stands for 'frequent to infrequent', ALL$_{I \to F}$ stands for 'infrequent to frequent', ALL$_{S \to L}$ stands for 'small to large' and ALL$_{L \to S}$ stands for 'large to small'.

The *parameter* of the above problems is the size of DB. One can easily construct examples for which the number of frequent connected subgraphs is exponential in this parameter. Thus, in general, the set of all frequent connected subgraphs cannot be computed in time polynomial only in the size of DB. Since this is a common feature of many listing problems, the following problem classes are usually considered in the literature (see, e.g., [6]). For some input I, let O be the output set of some finite cardinality N. Then the elements of O, say $o_1, \ldots, o_N$, are listed with:

- *polynomial delay* if the time before printing o_1, the time between printing o_i and o_{i+1} for every $i = 1, \ldots, N - 1$, and the time between printing o_N and the termination is bounded by a polynomial of the size of I,
- *incremental polynomial time* if o_1 is printed with polynomial delay, the time between printing o_i and o_{i+1} for every $i = 1, \ldots, N-1$ (resp. the time between printing o_N and the termination) is bounded by a polynomial of the combined size of I and the set $\{o_1, \ldots, o_i\}$ (resp. O),
- *output polynomial time* (or *polynomial total time*) if O is printed in time polynomial in the combined size of I and the *entire* output O.

Clearly, polynomial delay implies incremental polynomial time, which, in turn, implies output polynomial time. Furthermore, in contrast to incremental polynomial time, the delay of an output polynomial time algorithm may be exponential in the size of the input even before printing the first element of the output.

Example 1. Let us have graphs $G_1 = (\{1, 2, 3\}, \{\{1, 2\}, \{2, 3\}, \{3, 1\}\})$ and $G_2 = (\{1, 2, 3, 4\}, \{\{1, 2\}, \{2, 3\}, \{3, 4\}\})$, $DB = \{G_1, G_2\}$ and let $t = 2$. A solution of the FCGM problem is

$$H_1 = (\{1\}, \{\}), OCC_1 = \{1, 2\}$$
$$H_2 = (\{1, 2\}, \{\{1, 2\}\}), OCC_2 = \{1, 2\}$$
$$H_3 = (\{1, 2, 3\}, \{\{1, 2\}, \{2, 3\}\}), OCC_3 = \{1, 2\}$$

where OCC_i denotes the occurrences of the graph H_i. A solution of the problem $\text{ALL}_{F \to I}$ is

$$H_1 = (\{1\}, \{\}), OCC_1 = \{1, 2\}$$
$$H_2 = (\{1, 2\}, \{\{1, 2\}\}), OCC_2 = \{1, 2\}$$
$$H_3 = (\{1, 2, 3\}, \{\{1, 2\}, \{2, 3\}\}), OCC_3 = \{1, 2\}$$
$$H_4 = (\{1, 2, 3\}, \{\{1, 2\}, \{2, 3\}, \{3, 1\}\}), OCC_4 = \{1\}$$
$$H_5 = (\{1, 2, 3, 4\}, \{\{1, 2\}, \{2, 3\}, \{3, 4\}\}), OCC_4 = \{2\}$$

Remark 1. There is an incremental-polynomial-time algorithm for the FCGM (FCIGM) problem if and only if there is an incremental-polynomial time algorithm for $\text{ALL}_{F \to I}$ with (induced) subgraph isomorphism as a pattern matching operator.

In general, we will see in the next sections that the different ordered mining problems possess different computational complexities under standard complexity theoretic assumptions.

4   Mining All (Induced) Subgraphs

4.1   Negative Results

In this section, we provide several negative results regarding complexity of some of the enumeration problems considered in this paper. The first theorem connects the hardness of the frequent subgraph enumeration problem to fixed-parameter tractability of the pattern matching operator (subgraph isomorphism or induced subgraph isomorphism).

Theorem 1. *Let $\mathcal{G}$ be a class of graphs. Let $\preceq$ be either subgraph isomorphism or induced subgraph isomorphism. If deciding $H \preceq G$ for connected $G, H \in \mathcal{G}$ is not fixed-parameter tractable with the parameter $|H|$ then there is no output-polynomial-time algorithm for enumerating frequent connected graphs from databases consisting of graphs from $\mathcal{G}$.*

Proof. Let us suppose that there is an algorithm for mining frequent connected graphs from databases of graphs from $\mathcal{G}$ which runs in output-polynomial time. Let $G, H \in \mathcal{G}$. We show that then it is always possible to decide whether $H \preceq G$ holds in time $f(|H|) \cdot |G|^{O(1)}$. If $|H| > |G|$ then $H \npreceq G$ and we can finish. If $|H| \leq |G|$, we set $DB = \{H, G\}$ and we let the pattern mining algorithm run on DB with minimum frequency $t = 2$. Since there are at most $2^{|E(H)|}$ connected subgraphs of H, the mining algorithm will produce an output of length at most $|H| \cdot 2^{|E(H)|}$ in time $\mathsf{poly}\left(|H| + |G|, |H| \cdot 2^{|E(H)|}\right)$. If the output contains a frequent graph F such that $|V(F)| = |V(H)|$ and $|E(F)| = |E(H)|$ (such a graph F must be isomorphic to H), we return *true*. Otherwise, we return *false*. Now, if we return true then $H \preceq G$ must hold because if a graph F isomorphic to H is frequent, it must hold $F \preceq G$ and therefore also $H \preceq G$. Similarly, if we return false then there is no frequent graph F isomorphic to H and therefore $H \npreceq G$. This all put together runs in time $f(|H|) \cdot |G|^{O(1)}$ where f is a computable function. However, then the just described procedure would give us an algorithm for the pattern matching operator which would be fixed-parameter tractable with the parameter $|H|$ which is a contradiction.   □

From the proof of the above theorem, we can obtain the following corollary[2].

---

[2] The hardness for the $\mathsf{ALL}_{F \to I}$ problem follows from Theorem 1 together with Remark 1, whereas hardness of the $\mathsf{ALL}_{S \to L}$ problem follows from a simple modification of the proof of Theorem 1 where we use a hypothetic incr.-poly.-time algorithm for solving the $\mathsf{ALL}_{S \to L}$ problem and stop it after printing the first graph with more edges than H (or more vertices than H in the case of induced subgraph mining).

Corollary 1. *Let $\mathcal{G}$ be a class of graphs. Let $\preceq$ be either subgraph isomorphism or induced subgraph isomorphism. If deciding $H \preceq G$ for connected $G, H \in \mathcal{G}$ is not fixed-parameter tractable with the parameter $|H|$ then $\mathsf{ALL}_{F \to I}$ and $\mathsf{ALL}_{S \to L}$ cannot be solved in incremental polynomial time.*

However, there are also graph classes with FPT subgraph isomorphism, e.g. planar graphs [9], for which $\mathsf{ALL}_{F \to I}$ cannot be solved in incr.-poly. time[3].

Theorem 2. *The problem $\mathsf{ALL}_{F \to I}$ cannot be solved in incremental polynomial time for the class $\mathcal{G}$ of planar graphs and subgraph isomorphism as pattern matching operator.*

Proof. We can use NP-hardness of Hamiltonian cycle problem [2], similarly as [5]. We construct a database consisting of a given graph G and a cycle C on $|V(G)|$ vertices. A graph isomorphic to C with frequency 2 (recall that frequency is given implicitly by the printed occurrences) is output by the mining algorithm among the first $|V(G)| + 1$ graphs if and only if G contains a Hamiltonian cycle. It is easy to see that we could then use an incremental-polynomial time algorithm for the $\mathsf{ALL}_{F \to I}$ problem to solve the Hamitonian cycle problem. Therefore there is no incremental-polynomial time algorithm for the $\mathsf{ALL}_{F \to I}$ problem (unless P=NP).   □

This theorem is interesting because in Sect. 4.2, we will see that the problem $\mathsf{ALL}_{L \to S}$ can be solved with polynomial delay for planar graphs.

Even stronger negative result can be obtained for the problem $\mathsf{ALL}_{I \to F}$.

Theorem 3. *Let $\mathcal{G}$ be a class of graphs. Let $\preceq$ be either subgraph isomorphism or induced subgraph isomorphism. If deciding $H \preceq G$ for connected $G, H \in \mathcal{G}$ is NP-hard then $\mathsf{ALL}_{I \to F}$ cannot be solved in incremental polynomial time (unless P = NP).*

Proof. Let H and G be graphs from $\mathcal{G}$. We will show how to use an incremental-polynomial-time algorithm for the problem $\mathsf{ALL}_{I \to F}$ to decide whether $H \preceq G$ in polynomial time. We construct a database of graphs $DB = \{H, G, G\}$ and let the algorithm for the problem $\mathsf{ALL}_{I \to F}$ run until it outputs a graph and its occurrences (implicitly giving us also the frequency) and then we stop it (it follows from definition of incremental-polynomial time that this will run only for time polynomial in the sizes of H and G). It is easy to see that the output graph has frequency 1 if and only if $H \not\preceq G$. Thus, if the frequency of the output graph is 1 we return 'not (induced) subgraph isomorphic' and if the frequency of the output graph is greater than 1 then we return '(induced) subgraph isomorphic'. Therefore if deciding $H \preceq G$ is NP-hard for graphs from $\mathcal{G}$ there cannot be an incremental-polynomial-time algorithm for the problem $\mathsf{ALL}_{I \to F}$ (unless P = NP).   □

Using the fact that (induced) subgraph isomorphism is NP-complete even for bounded-treewidth graphs [10] and planar graphs [9], we can obtain the following.

---

[3] The complexity of the $\mathsf{ALL}_{S \to L}$ problem in these cases remains an interesting open problem.

Corollary 2. *The problem* ALL$_{I \to F}$ *cannot be solved in incremental polynomial time for the class of planar graphs and for the class of bounded-treewidth graphs.*

Note that Theorem 1 cannot be made as strong as Theorem 3 (i.e. showing that ALL$_{F \to I}$ cannot be solved in incremental-polynomial time if the pattern matching operator is NP-hard) because the results of Horváth and Ramon from [5] demonstrate that even if the pattern matching operator is NP-hard there can be an incremental-polynomial-time algorithm for mining frequent subgraphs. Theorem 3 shows that we cannot expect such a result for mining *infrequent* subgraphs (i.e. subgraphs with frequency below a threshold).

4.2   Positive Results for ALL$_{F \to I}$ and ALL$_{S \to L}$

Before presenting our new results for ALL$_{L \to S}$ in the next section, we note that there exists the following positive result for frequent graph mining from bounded-treewidth graphs, which was presented in [4,5].

Theorem 4 (Horváth and Ramon [5], Horváth, Otaki and Ramon [4]). *The FCGM and FCIGM problems can be solved in incremental-polynomial time for the class of bounded-treewidth graphs.*

This result directly translates to a positive result for the problem ALL$_{F \to I}$ summarized in the following corollary (recall that we have shown in the previous section that ALL$_{I \to F}$ cannot be solved in incremental-polynomial time for bounded-tree-width graphs) and to a result for the problem ALL$_{S \to L}$ (this other result follows from the fact that the respective algorithms are level-wise).

Corollary 3. *The problems* ALL$_{F \to I}$ *and* ALL$_{S \to L}$ *can be solved in incremental-polynomial time for the class of bounded-treewidth graphs.*

4.3   Positive Results for ALL$_{L \to S}$

In this section, we describe an algorithm called LARGERTOSMALLER (Algorithm 1) which, when given a class of graphs $\mathcal{G}$ in which isomorphism can be decided in polynomial time, solves the problem ALL$_{L \to S}$ in incremental-polynomial time, or with polynomial delay if $\mathcal{G}$ also admits a polynomial-time canonization. The main employed trick is the observation that for the problem ALL$_{L \to S}$ it is not necessary to use subgraph isomorphism explicitly for computing occurrences.

The algorithm maintains a data structure ALL storing key-value pairs where keys are graphs and values are sets of IDs[4] of graphs in which the given key graph is contained either as a subgraph or as an induced subgraph (depending on whether we are mining subgraphs or induced subgraphs). The data structure provides four functions/procedures: ADD(K, OCC, ALL), GET(K, ALL), KEYS(n, ALL), and DELETE(n, ALL).

---

[4] Here, IDs are just some identifiers given to the database graphs.

The procedure ADD(K, OCC, ALL) adds the IDs contained in OCC to the set associated with a key contained in ALL which is isomorphic to K or, if no such key is contained in ALL, the procedure stores K in ALL and associates OCC with it. If we restrict attention to graphs from a class $\mathcal{G}$ for which a polynomial-time canonization function running in $O(p(|H|))$ exists (where p is a polynomial) then the procedure ADD(K, OCC, ALL) can be implemented to run in time $O(p(|K|))$ (we can just store the key graphs as canonical strings, therefore a hashtable with constant-time methods for finding and adding values by their keys can be used). If a polynomial-time canonization function does not exist but graph isomorphism can be decided in time $O(p_{iso}(|K|))$ where p_{iso} is a polynomial then the procedure ADD(K, OCC, ALL) can be implemented to run in time $O(|\{K' \in KEYS(ALL) : |V(K')| = |V(K)| \text{ and } |E(K')| = |E(K)|\}| \cdot p_{iso}(|K|))$.

The function GET(K, ALL) returns all IDs associated with a key isomorphic to K. The exactly same considerations as for the procedure ADD apply also for this function.

The function KEYS(n, ALL) returns a pointer to a linked list containing all key graphs stored in ALL which have size n. Since the data structure ALL does not allow deletion of individual keys, it is easy to maintain such a linked list[5].

Finally, the procedure DELETE(n, ALL) removes the pointer[6] to the linked list containing all key graphs of order n stored in ALL.

The algorithm LARGERTOSMALLER fills in the data structure ALL, starting with the largest graphs and proceeding to the smaller ones. When it processes a graph H, it first prints it and the IDs of the graphs associated to it in the data structure ALL, and then it calls the function REFINE which returns all connected subgraphs H' of H which can be obtained from H by removing an edge or an edge and its incedent vertex of degree one, in the case of subgraph mining, or just by removing a vertex and all its incident edges, in the case of induced subgraph mining. It then associates all occurrences of the graph H with the graphs H' in the datastructure ALL using the procedure ADD. Since the same graph H' may be produced from different graphs H, the occurrences of H' accumulate and we can prove that when a graph H is printed, the data structure ALL already contains all IDs of graphs in which H is contained.

Theorem 5. *Let $\mathcal{G}$ be a hereditary class of graphs with isomorphism decidable in polynomial time. Given a database DB of connected graphs from $\mathcal{G}$, the algorithm LARGERTOSMALLER solves the problem $ALL_{L \to S}$ in incremental polynomial time. If the graphs from $\mathcal{G}$ also admit a poly-time canonization then the algorithm LARGERTOSMALLER solves the problem $ALL_{L \to S}$ with polynomial delay.*

---

[5] The reason why the function KEYS does not just return all the key graphs but rather a pointer to the linked list is that if it did otherwise, Algorithm 1 could never run with polynomial delay.

[6] Note that we just remove the pointer and do not actually "free" the memory occupied by the graphs. For the practical implementation, we used a programming language with a garbage collector.

Algorithm 1. LARGERTOSMALLER

Require: database DB of connected transaction graphs
Ensure: all connected (induced) subgraphs and their occurrences

1: **let** ALL be a data structure for storing graphs and their occurrences (as described in the main text).
2: **for** $G \in DB$ **do**
3:     ADD$(G, \{\text{ID}(G)\}, ALL)$
4: **endfor**
5: **let** $m := \max_{G \in DB} |E(G)|$ ($m := \max_{G \in DB} |V(G)|$ for induced subgraph mining).
6: **for** $(l := m; l > 0; \; l := l - 1)$ **do**
7:     **for** $H \in \text{KEYS}(l, ALL)$ **do**
8:         $OCC \leftarrow \text{GET}(H, ALL)$
9:         PRINT(H, OCC)
10:        **for** $H' \in \text{REFINE}(H)$ **do**
11:            **if** H' is connected **then**
12:                ADD(H', OCC, ALL)
13:            **endif**
14:        **endfor**
15:    **endfor**
16:    DELETE(l, ALL)
17: **endfor**

Proof. First, we show that the algorithm prints every (induced) subgraph of the graphs in DB. Let us assume, for contradiction, that this is not the case and let G^* be a maximal connected graph which is a (induced) subgraph of a graph in the database and such that no graph isomorphic to it is printed by the algorithm. It is easy to verify that such a graph G^* cannot be isomorphic to any graph in DB. Since G^* is not isomorphic to a graph from DB and since it is a maximal graph not printed, there must be a supergraph G' of G^* such that $|E(G^*)|+1 = |E(G')|$ in the case of subgraph isomorphism ($|V(G^*)|+1 = |V(G')|$ in the case of induced subgraph isomorphism, respectively) and such that a graph isomorphic to it is printed by the algorithm. However, if such a graph was printed then a graph isomorphic to G^* would have to be in $\text{REFINE}(G')$ and would have to be printed eventually, which is a contradiction.

Second, we show that the occurrences printed with each graph are correct. First, if a printed graph G does not have any strict supergraph in the database then it must be equivalent to one or more database graphs. However, it is easy to check by simple inspection of the algorithm that the occurrences must be correct in this case. For the rest of the printed graphs (i.e. graphs which have strict supergraphs in the databse), let us assume, for contradiction, that G^* is a maximal graph printed by the algorithm for which the printed occurrences are not correct, i.e. either there is an ID of a database graph printed for G^* of which G^* is not an (induced) subgraph or there is an ID of a database graph not printed for G^* of which G^* is actually an (induced) subgraph. *(False occurrence:)* If there is an ID of a database graph of which G^* is not a (induced) subgraph then at least one of the graphs from which G^* can be obtained by refinement must have an ID associated which it should not have. But then G^* could not be maximal graph with this property, which is a contradiction. *(Missing occurrence:)* If there is a missing ID of a database graph of which G^* is a (induced) subgraph then one of the following must be true: (i) G^* is isomorphic to the database graph but then it is easily seen by inspection of the algorithm that the ID of this graph cannot be missing from the occurrences of G^*, (ii) there is a strict supergraph G' of G^* which is (induced) subgraph isomorphic to the database graph and which is not printed (and therefore its occurrences are not added to the data structure ALL), but this is not possible as the first part of the proof shows, (iii) there is a strict supergraph G' of G^* which is (induced) subgraph isomorphic to the database graph and the respective ID was not associated to it, but then G^* could not be a maximal graph with this property. Thus, we have a contradiction.

Third, we show that if there is a polynomial-time isomorphism algorithm then the algorithm LARGERTOSMALLER runs in incremental-polynomial time. First, notice that the first for-loop takes only polynomial time in the size of the database. We can see easily that the time before printing the first graph is also bounded by a polynomial in the size of the database. Next, the for-loop on line 10 is repeated at most $|H|$-times for any graph H. Adding a graph H' to the data structure ALL or getting occurrences of a graph H' from the data structure ALL takes time polynomial in the number of graphs already stored in it and the size of the graph being stored (which is discussed in the main text). The number of graphs already stored in the data structure ALL is bounded by $P \cdot M$ where P is the number of already printed graphs and M is the maximum size of a graph in the database. Thus, we have that the time between printing two consecutive graphs is bounded by a polynomial in the size of the database and in the number of already printed graphs, i.e. the algorithm runs in incremental polynomial time.

Fourth, we can show using essentially the same reasoning that if there is a polynomial-time graph canonization algorithm then the algorithm runs with polynomial delay. □

Using the results on complexity of graph canonization for planar [12] and bounded-treewidth graphs [7], we can get the following corollary.

Corollary 4. *The problem* $\text{ALL}_{L \to S}$ *can be solved with polynomial delay for the classes of planar and bounded-treewidth graphs.*

In fact, there are many other classes of graphs for which graph isomorphism is known to be decidable in polynomial time, e.g. graphs of bounded degree [8]. One can easily use the theorems presented in this section for such classes too as long as they are hereditary, as is the case for bounded-degree graphs.

4.4   Other Negative Results

The following theorem asserts that the results for the problem $\text{ALL}_{L \to S}$ are essentially optimal in the sense that existence of a polynomial-time algorithm for graph isomorphism is both sufficient and necessary for existence of an incremental-polynomial-time algorithm for the problem $\mathsf{ALL}_{L \to S}$.

Theorem 6. *The problem $\mathsf{ALL}_{L \to S}$ can be solved in incremental-polynomial time for graphs from a hereditary class $\mathcal{G}$ if and only if graph isomorphism can be decided in polynomial-time for graphs from $\mathcal{G}$.*

Proof. ($\Rightarrow$) The proof idea is similar to the idea of the proof of Theorem 3. Let H and G be graphs from $\mathcal{G}$. We will show how to use an incremental-polynomial-time algorithm for the problem $\mathsf{ALL}_{L \to S}$ to decide whether H and G are isomorphic in polynomial time. If $|V(H)| \neq |V(G)|$ or $|E(H)| \neq |E(G)|$, we return 'not isomorphic'. Otherwise, we construct a database of graphs $DB = \{H, G\}$ and let the algorithm for the problem $\mathsf{ALL}_{L \to S}$ run until it outputs a graph and its occurrences (implicitly giving us also the frequency) and then we stop it (it follows from definition of incremental-polynomial time that this will run only for time polynomial in the sizes of H and G). Then the output graph has frequency 1 if and only if H and G are not isomorphic. So, if the frequency of the output graph is 1 we return 'not isomorphic' and if the frequency of the output graph is greater than 1 then we return 'isomorphic'. This gives us an algorithm for deciding isomorphism of graphs which runs in polynomial time.

($\Leftarrow$) This direction is explicitly shown in Theorem 5. □

The next theorem indicates that $\mathsf{ALL}_{L \to S}$ is the simplest (complexity-wise) from the enumeration problems considered in this paper because e.g. the problems $\mathsf{ALL}_{F \to I}$ and $\mathsf{ALL}_{I \to F}$ may be unsolvable in incremental-polynomial time even if a polynomial-time graph isomorphism algorithm existed.

Theorem 7. *If $\mathsf{GI} \in \mathsf{P}$ and $\mathsf{P} \neq \mathsf{NP}$ was true then there would be an incremental-polynomial-time algorithm for the problem $\mathsf{ALL}_{L \to S}$ for the class $\mathcal{G}$ of all graphs but no incremental-polynomial-time algorithm for the problems $\mathsf{ALL}_{I \to F}$ and $\mathsf{ALL}_{F \to I}$ for the class of all graphs.*

Proof. The positive result for the problem $\mathsf{ALL}_{L \to S}$ follows from Theorem 5. The hardness of FCGM and FCIGM has been shown in [5] and in [4] using reductions from Hamiltonian cycle problem (for mining under subgraph isomorphism) and from maximum clique problem (for mining under induced subgraph isomorphism), from which the hardness result for the $\mathsf{ALL}_{F \to I}$ problem follows. The hardness of the $\mathsf{ALL}_{I \to F}$ problem follows from Theorem 3. □

For now, we leave open the question of complexity of the $\mathsf{ALL}_{S \to L}$ problem conditioned only on the pattern matching operator not being in P. Theorem 1 asserts that solving this problem in incremental polynomial time is not possible if the pattern matching operator is not fixed-parameter tractable with the parameter being the size of the pattern graph, which is a widely believed conjecture.

5 Mining Under Homeomorphism and Minor Embedding

Many of the results presented in this paper may be generalized to mining with other important pattern matching operators: (induced) homeomorphism and

(induced) minor embedding. In this section, we briefly discuss these generalizations. Here, we only consider mining from unlabeled graphs because there is
no generally agreed-upon definition of homeomorphism or minor embedding of
labeled graphs for pattern mining[7].

The ideas from Theorem 1 are not relevant for mining under minor embedding
or homeomorphism because minor embedding and homeomorphism are fixed-
parameter tractable with the parameter being the size of the pattern graph as
shown in [3,11]. However, it can be used together with the following theorem
to show hardness of the $\text{ALL}_{S \to L}$ problem under induced homeomorphism and
induced minor embedding.

Theorem 8. *Deciding induced homeomorphism or induced minor embedding*
$G_1 \preccurlyeq G_2$ *is not fixed-parameter tractable with the size of G_1 as the parame-*
ter (unless FPT = W[1]).

Proof. This can be shown by reduction from the k-independent set problem
parameterized by k which follows from the following simple observation. Let G
be a graph. G contains an independent set of size k if and only if $H \preccurlyeq G$ where
H is a graph consisting of k isolated vertices. Notice that this theorem holds
also when we restrict G_1 and G_2 to be connected graphs. The basic idea of the
proof is then the same. The reduction from k-independent set problem is then
as follows. We create a new graph G' by taking the graph G from the proof,
adding a new vertex and connecting it to all vertices of G. Instead of taking H
as a set of k isolated vertices we let H be a star on $k + 1$ vertices (i.e. a tree in
which all vertices are connected to one vertex v). We then again have $H \preccurlyeq G'$ if
and only if G contains an independent set of size k. $\square$

Ideas analogical to those from Theorems 1, 2, 3 and 8 can be used to obtain
the following negative results for mining under (induced) homeomorphism and
(induced) minor embedding.

Theorem 9. *The problem $\text{ALL}_{S \to L}$ under induced homeomorphism or induced*
minor embedding cannot be solved in incremental-polynomial time for the class
of all graphs (unless FPT = W[1]). The problems $\text{ALL}_{F \to I}$ under (induced)
homeomorphism or (induced) minor embedding cannot be solved in incremental-
polynomial time for the class of all graphs (unless P = NP). The problem
$\text{ALL}_{I \to F}$ under (induced) homeomorphism or (induced) minor embedding can-
not be solved in incremental-polynomial time for the class of bounded-treewidth
graphs (unless P = NP).

The positive results from Sect. 4.2 may be adapted for mining under (induced)
homeomorphism and (induced) minor embedding as follows. We can essentially

---

[7] For homeomorphism, for instance, we could allow the subdivided edges to have
different labels or, to the contrary, we could require them all to have the same label
etc. Then another question could be how we should treat labels of vertices etc. While
these considerations are interesting even for practice, they are out of the scope of
this paper.

use the algorithm LARGERTOSMALLER as is, but we need to modify the procedure REFINE(H) (the modified algorithm will be denoted as LARGERTO-SMALLER* to avoid confusion). For mining under minor embedding, when given a graph H, the procedure REFINE should return all graphs H' which can be obtained from H by removing an edge or by contracting an edge and removing loops and multiple edges thus produced. For mining under homeomorphism, it should return all graphs H' which can be obtained from H by removing an edge or by contracting an edge incident to a vertex of degree 2. For mining under induced minor embedding, it should return all graphs H' which can be obtained from H by removing a vertex and all its incident edges or by contracting an edge and removing loops and multiple edges thus produced. Finally, for mining under induced homeomorphism, it should return all graphs H' which can be obtained from H by removing a vertex and all its incident edges or by contracting an edge incident to a vertex of degree 2.

Theorem 10. *Let $\mathcal{G}$ be a class of graphs closed under formation of minors admitting a polynomial-time isomorphism algorithm and let the pattern matching operator $\preccurlyeq$ be either (induced) homeomorphism of (induced) minor embedding. Given a database DB of connected graphs from $\mathcal{G}$, the algorithm LARGERTO-SMALLER* solves the problem $\mathrm{ALL}_{L \to S}$ in incremental polynomial time. If the graphs from $\mathcal{G}$ also admit a poly-time canonization then the algorithm LARGER-TOSMALLER* solves the problem $\mathrm{ALL}_{L \to S}$ with polynomial delay.*

Proof. (Sketch). We can essentially use the reasoning from the proof of Theorem 5. We only need to notice that $\preccurlyeq$ is transitive and that if $H \preccurlyeq G$ then a graph isomorphic to H can be obtained from G by a repeated application of the procedure REFINE. For instance, to show that all graphs homeomorphic to at least one database graph will be printed eventually, we can reason as follows. We can show using Theorem 5 that every graph which is subgraph isomorphic to at least one database graph must also be printed by the algorithm LARGERTOSMALLER*. If a graph H is homeomorphic to a database graph G then G has a subgraph G', which must be printed at some point, which is isomorphic to a graph which can be obtained from H by subdividing its edges. But this also means that H can be obtained from G' by repeatedly contracting some of its edges incident to a vertex of degree 2 and removing the loops produced by this process. Thus, any graph homeomorphic to at least one of the database graphs will be printed eventually. It is also not difficult to see that all graphs produced by the refinement function REFINE must be homeomorphic to the graph being refined. Since homeomorphism is transitive any of the produced graphs will be homeomorphic to at least one of the database graphs. Finally, to show that the occurrences printed for every output graph are correct, we can repeat the reasoning from Theorem 5 (which we omit here due to space constraints). □

Using the fact that bounded-treewidth and planar graphs are closed under formation of graph minors and that they also admit a polynomial-time isomorphism algorithm, we can obtain the following corollary.

Corollary 5. *The problem* ALL$_{L \to S}$ *under (induced) homeomorphism or (induced) minor embedding can be solved with polynomial delay for the classes of planar and bounded-treewidth graphs.*

6   Conclusions and Future Work

In this paper, we have shown how different orders in which graphs are enumerated affect computational complexity of the mining problem. We have presented several negative results. We also described a positive result which shows that it is possible to mine all graphs from a database of bounded-treewidth or planar graphs with polynomial delay under any of the following six different pattern matching operators: (induced) isomorphism, (induced) homeomorphism and (induced) minor embedding. Here, by mining all graphs, we mean enumerating all graphs which have frequency at least one and printing their occurrences in the database. This result holds despite the fact that deciding any of the six pattern matching operators is NP-hard. However, since the positive result depends heavily on mining all graphs and not just the frequent ones, the question whether frequent graph mining is achievable with polynomial delay for an NP-hard pattern matching operator, remains open. In fact, as we have shown, e.g. for planar graphs the latter problem of mining frequent graphs cannot be done even in incremental polynomial time, whereas mining all graphs can be done with polynomial delay.

Lastly, it is worth noting that when we performed preliminary experiments with a simple implementation of the algorithm for the ALL$_{L \to S}$ problem, we were able to mine completely about 70 % molecules from NCI GI dataset consisting of approximately three and half thousand organic molecules. This suggests that the techniques presented in this paper might also lead to development of practical graph mining algorithms. For instance, it would not be difficult to obtain a graph mining algorithm, having similar positive complexity guarantees as the LARG-ERTOSMALLER algorithm, for mining graphs of bounded radius (from database graphs of arbitrary radius) or with constraints on minimum vertex degree[8] etc. Exploring these and similar ideas is left for future work.

Acknowledgement. This work has been supported by ERC Starting Grant 240186 "MiGraNT: Mining Graphs and Networks, a Theory-based approach". The first author is supported by a grant from the Leverhulme Trust (RPG-2014-164).

---

[8] The latter constraint on degrees would not probably be very relevant for bounded-treewidth graphs or planar graphs. This is because any graph of treewidth k must always have at least one vertex of degree at most k and because any planar graph must always have at least one vertex of degree at most 5. However, isomorphism algorithms are extremely fast in practice, despite not being polynomial in the worst case, so the algorithm that we would obtain for general graphs could still be practical.

References

1. Flum, J., Grohe, M.: Parameterized Complexity Theory. Springer, Berlin (2006)
2. Garey, M.R., Johnson, D.S., Tarjan, R.E.: The planar hamiltonian circuit problem is np-complete. SIAM J. Comput. **5**(4), 704–714 (1976)
3. Grohe, M., Marx, D.: Structure theorem and isomorphism test for graphs with excluded topological subgraphs. CoRR, abs/1111.1109 (2011)
4. Horváth, T., Otaki, K., Ramon, J.: Efficient frequent connected induced subgraph mining in graphs of bounded tree-width. In: Blockeel, H., Kersting, K., Nijssen, S., Železný, F. (eds.) ECML PKDD 2013, Part I. LNCS, vol. 8188, pp. 622–637. Springer, Heidelberg (2013)
5. Horváth, T., Ramon, J.: Efficient frequent connected subgraph mining in graphs of bounded tree-width. Theoret. Comput. Sci. **411**(31–33), 2784–2797 (2010)
6. Johnson, D.S., Yannakakis, M., Papadimitriou, C.H.: On generating all maximal independent sets. Inf. Process. Lett. **27**(3), 119–123 (1988)
7. Lokshtanov, D., Pilipczuk, M., Pilipczuk, M., Saurabh, S.: Fixed-parameter tractable canonization and isomorphism test for graphs of bounded treewidth. In: 55th IEEE Annual Symposium on Foundations of Computer Science, FOCS 2014, pp. 186–195 (2014)
8. Luks, E.M.: Isomorphism of graphs of bounded valence can be tested in polynomial time. J. Comput. Syst. Sci. **25**(1), 42–65 (1982)
9. Marx, D., Pilipczuk, M.: Everything you always wanted to know about the parameterized complexity of subgraph isomorphism (but were afraid to ask). In: 31st International Symposium on Theoretical Aspects of Computer Science, STACS 2014, pp. 542–553 (2014)
10. Matoušek, J., Thomas, R.: On the complexity of finding iso- and other morphisms for partial k-trees. Discrete Math. **108**(1–3), 343–364 (1992)
11. Robertson, N., Seymour, P.D.: Graph minors. XIII. the disjoint paths problem. J. Comb. Theory Ser. B **63**(1), 65–110 (1995)
12. Torán, J., Wagner, F.: The complexity of planar graph isomorphism. Bull. EATCS **97**, 60–82 (2009)

Processing Markov Logic Networks with GPUs: Accelerating Network Grounding

Carlos Alberto Martínez-Angeles[1(✉)], Inês Dutra[2], Vítor Santos Costa[2], and Jorge Buenabad-Chávez[1]

[1] Departamento de Computación, CINVESTAV-IPN, Av. Instituto Politécnico Nacional 2508, 07360 México D.F., Mexico
camartinez@cinvestav.mx, jbuenabad@cs.cinvestav.mx
[2] Departamento de Ciência de Computadores, CRACS INESC-TEC LA and Universidade do Porto, Rua do Campo Alegre 1021, 4169-007 Porto, Portugal
{ines,vsc}@dcc.fc.up.pt

Abstract. Markov Logic is an expressive and widely used knowledge representation formalism that combines logic and probabilities, providing a powerful framework for inference and learning tasks. Most Markov Logic implementations perform inference by transforming the logic representation into a set of weighted propositional formulae that encode a Markov network, the ground Markov network. Probabilistic inference is then performed over the grounded network.

Constructing, simplifying, and evaluating the network are the main steps of the inference phase. As the size of a Markov network can grow rather quickly, Markov Logic Network (MLN) inference can become very expensive, motivating a rich vein of research on the optimization of MLN performance. We claim that parallelism can have a large role on this task. Namely, we demonstrate that widely available Graphics Processing Units (GPUs) can be used to improve the performance of a state-of-the-art MLN system, Tuffy, with minimal changes. Indeed, comparing the performance of our GPU-based system, TuGPU, to that of the Alchemy, Tuffy and RockIt systems on three widely used applications shows that TuGPU is up to 15x times faster than the other systems.

Keywords: Statistical relational learning · Markov logic · Markov logic networks · Datalog · Parallel computing · GPUs

1 Introduction

Statistical relational learning (SRL) integrates statistical reasoning, machine learning and relational representations. SRL systems rely on a first-order logic language to represent the structure and relationships in the data, and on graphical models to address noisy and incomplete information. Various SRL frameworks have been proposed, Stochastic Logic Programs (SLP), Probabilistic Relational Models (PRM), PRISM, Bayesian Logic Programs, ProbLog, CLP($\mathcal{BN}$), PFL, and Markov Logic [13,30].

© Springer International Publishing Switzerland 2016
K. Inoue et al. (Eds.): ILP 2015, LNAI 9575, pp. 122–136, 2016.
DOI: 10.1007/978-3-319-40566-7_9

The last few years have seen significant progress in models that can represent and learn from complex data. One important such model is Markov logic, "a language that combines first-order logic and Markov networks. A knowledge base in Markov logic is a set of first-order [logic] formulas with weights" [11]. Markov Logic thus builds upon logic and probabilities. The logical foundation of Markov Logic provides the ability to use first-order logic formulas to establish *soft constraints* over *worlds*, or interpretations. Worlds that violate a formula are less likely to be true, but still possible. In contrast, formulas in standard first-order logic are *hard constraints*: a world that falsifies a formula is not possible. Worlds that violate a formula can be possible in Markov Logic because worlds are an assignment to a set of random variables, and follow a probability distribution. The distribution is obtained by identifying each ground atom as a random variable, and each grounded formula as a clique in a factor graph. This ground network thus forms a Markov Random Field (MRF) [17].

Markov logic systems address two major tasks [31]: inference and learning. In *inference*, we receive an MLN model M and a set of observations, or evidence E, and we want to ask questions about the unobserved variables. Typical queries are:

- *probability estimation queries*: one wants to find out the probability of an atom given the evidence E. A typical example would be "What is the probability of rain in Kobe and Kyoto, given that it is raining in Tokyo and Nagoya, but sunny in Fukuoka and Okinawa". Notice that MLNs naturally allow for collective inference, that is, we can ask for all the different cities in a single query.
- Maximum a posteriori (MAP) or *most likely world queries*: one wants to find out what is the most likely set of values for the variables of interest. From our example above, instead of outputting probabilities, the model would output the places where it is more likely to rain.

A large number of inference techniques have been developed for MLNs. Most of them operate on the ground network, that is, given the query and the observed data, they enumerate all relevant atoms and then use statistical inference on the resulting network. They then search for the set of grounded clauses that maximize the sum of the satisfied clauses weights.

The second task, *learning*, is about constructing the actual MLNs. Often the formulas of interest can be obtained from the experts, but it is still necessary to learn the weights. *Parameter learning* addresses this task. *Structure learning* goes further and tries to construct the actual model, by searching for relationships or important properties of the data.

Markov logic networks have been widely adopted. Applications include the Semantic Network Extractor (SNE) [19], a large scale system that can learn semantic networks from the Web; the work by We *et al.* to refine Wikipedia's Infobox Ontology [43]; and Riedel and Meza-Ruiz's work to carry out collective semantic role labelling [33], among others [11, p. 97].

Alchemy was the first widely available Markov Logic system [11]. It is still a reference in the field, as it includes a very large number of algorithms that address most MLN tasks. However, as it did not scale well to large real-world applications, several new implementations have been proposed [4,26,27,32].

Grounding is, arguably, the step that mostly affects performance of MLNs, preventing them from scaling to large applications. For large domains, we may need to ground a very large number of atoms, which can be quite time and space-consuming. Often (but not always) the solver algorithm converges in few iterations and grounding will dominate running time [26]. We claim that GPU processing can significantly expedite grounding, and that this can be done effectively with few changes to the state-of-the-art systems. To verify our hypothesis, we designed TuGPU, a Markov Logic system based on: Tuffy [26], YAP Prolog [35], and GPU-Datalog, a GPU-based engine that evaluates Datalog programs [22]. We compare the performance of TuGPU to that of Alchemy [11], Tuffy and RockIt [27], with three applications of different types: information extraction, entity resolution and relational classification. The performance of TuGPU is on par or better than the other systems for most applications.

This paper is organized as follows. Section 2 presents background on Markov logic and its implementation, Tuffy, Datalog, and GPUs. Section 3 presents the design and implementation of our TuGPU platform for Markov logic networks. Section 4 presents an experimental evaluation of our platform. In Sect. 5, we discuss about our system and other related systems. We conclude in Sect. 6.

2   Markov Logic, Tuffy, Datalog and GPUs

First-order (predicate) logic is widely used for knowledge representation and inference tasks. Datalog is a language based on first-order logic that was initially investigated as a data model for relational databases in the 80s [41,42]; recent applications include declarative networking, program analysis, and security [15]. Interest in Datalog has always stemmed from its ability to compute the transitive closure of relations through recursive queries which, in effect, turns relational databases into deductive databases. Relational Learning is the task of learning from databases, modelling relationships among data items from multiple tables (relations); Inductive Logic Programming (ILP) [9] is a popular relational learning approach that employs logic-based formalisms, often based on subsets of first-order logic such as Horn clauses.

Statistical Relational Learning (SRL), in the form of probabilistic inductive logic programming, extends logic-based approaches by combining relational learning and probabilistic models (e.g., graphical models such as Bayesian networks and Markov networks), in order to manage the uncertainty arising from noise and incomplete information which is typical of real-world applications. Markov logic networks (MLNs) are a very popular approach that combines first-order logic and Markov networks in a simple manner: a weight is attached to each first-order logic formula that represents how strong the formula is as a constraint in all possible worlds. MLNs use inference to answer queries of the form: "What is the probability that formula F_1 holds given that formula F_2 does?" [31].

2.1   Inference in Markov Logic

A Markov Logic network is a set of formulas with attached weights. Internally, the program is stored as a conjunction of clauses, where each clause is a disjunction of positive and negative atoms, as shown in the well-known *smokers* example, which determines the probability of people having cancer (Ca) based on who their friends (Fr) are and whether or not their friends smoke (Sm):

$$1.5 : \neg\text{Sm}(x) \vee \text{Ca}(x)$$
$$1.1 : \neg\text{Fr}(x, y) \vee \neg\text{Sm}(y) \vee \text{Sm}(x)$$
$$0.7 : \neg\text{Fr}(x, y) \vee \neg\text{Fr}(y, z) \vee \text{Fr}(x, z)$$

Each ground instance of a literal, say Ca(Anna) can be seen as a boolean random variable (RV). RVs in a clause form a clique, and the set of all cliques a hypergraph. Assuming the network includes N ground atoms and R rules or cliques, such that clique i has size k_i, the Markov property says that the joint probability over the hypergraph is a normalized sum of products:

$$P(a_1, \ldots, a_N) = \frac{1}{\mathcal{Z}} \prod_R e^{w_i \phi(a_{i1}, \ldots, a_{ik_i})}$$

Each RV can take two values (0 or 1), hence we have 2^N disjoint configurations. The partition function $\mathcal{Z} = \sum_{(a_1=0\ldots a_N=0)}^{(a_1=1\ldots a_N=1)} \prod_R e^{w_i \phi(a_{i1} \ldots a_{ik_i})}$ sums up all the different values and ensures that the total probabilities add up to one (1). Usually there is no closed form for $\mathcal{Z}$.

The boolean function ϕ is 1 if the clause i is true under this grounding, 0 otherwise. Thus, a false grounding contributes $e^0 = 1$ to the product, and a true grounding e^w: in other words, if $w = 1.5$, a world with that grounding is $e^{1.5}$, which is approximately 5 times more likely than a world whose grounding is false. As $wF \equiv -w\neg F$ (where F is a a clique i), we can always ensure that weights are positive or zero, hence the probability of a world where all constraints are soft is $0 < \frac{1}{\mathcal{Z}} < \frac{\prod e^{w_i}}{\mathcal{Z}} < 1$: strictly larger than zero and always less than one.

Inference is most often divided in two phases: *grounding* and *search*. Grounding is the process of assigning values to all free variables in each clause. While we can ground a clause by assigning all possible values to its variables, it is impractical even for small domains. There are several, more efficient alternatives that discard unnecessary groundings, such as lazy closure grounding and inference [28]. In a number of cases, one can obtain even better results by using *lifted inference*, that avoids grounding the program [38].

Next we focus on the most common inference task, Maximum a Posteriori (MAP), where we search for the most probable state of the world given the observed data or *evidence* E; that is, we search for an assignment of $a_1 \ldots a_N$ that maximizes $P(E|a_1 \ldots a_N) \propto P(a_1 \ldots a_N|E) = \frac{P(a_1 \ldots a_N, E)}{P(E)}$. Thus, we have:

$$\text{argmax}_{a_1 \ldots a_N} P(a_1 \ldots a_N|E) = \text{argmax}_{a_1 \ldots a_N} \frac{1}{P(E)\mathcal{Z}} \prod_R e^{w_i \phi(a_{i1} \ldots a_{ik_i})}$$
$$= \text{argmax}_{a_1 \ldots a_N} \prod_R e^{w_i \phi(a_{i1} \ldots a_{ik_i})}$$
$$= \text{argmax}_{a_1 \ldots a_N} \log \prod_R e^{w_i \phi(a_{i1} \ldots a_{ik_i})}$$
$$= \text{argmax}_{a_1 \ldots a_N} \sum_R w_i \phi(a_{i1} \ldots a_{ik_i})$$

$\mathcal{Z}$ and $P(E)$ are the same for every world, so they do not affect the optimization problem. Moreover, applying a monotonic function such as the logarithm will preserve the maximum, but enable us to work on a sum. Observing closely, the problem reduces to finding the maximal value of discrete function of boolean variables. Notice that if the coefficients w_i are positive, and the underlying boolean formula is satisfiable, an assignment that satisfies the model will be optimal. Thus, finding a solution to this problem requires solving the satisfiability problem with weights.

Several Markov logic systems use MaxWalkSAT [16] for its ability to solve hard problems with thousands of variables in a short time. MaxWalkSAT works by selecting an unsatisfied clause and switching the truth value of one of its atoms. The atom is chosen either randomly or to maximize the sum of the satisfied clause weights.

2.2 Optimizations

A ground MLN may quickly have thousands of boolean variables, making it hard to find even an approximate solution. Thus, it is important to start by simplifications of the system. Typically, one applies a combination of two techniques:

- *Elimination*: Consider a clause $\neg Sm(j) \vee \neg Sm(k)$, with evidence $\neg Sm(j)$ and query variable $Sm(k)$. The clause is always true, hence it does not affect the total score, and can be dropped.
- *Partitioning*: Consider $c_1 \equiv a \vee \neg b$ and $c_2 \equiv c \vee d$. If (a, b) is a solution to c_1, and (c, d) is a solution to c_2, then we have that (a, b, c, d) is a solution to the joint network. In practice this means the two sub-problems can be solved independently.

Most MLN systems apply these principles to reduce the search space. To speed up inference in large relational problems, *lazy grounding* takes the idea further and grounds as late as possible. The idea is to take advantage of the fact that most of their groundings are known to be trivial or false beforehand. The other approach, lifted inference, groups indistinguishable atoms together and treats them as a single unit, thus reducing the size of the network.

2.3 Learning

Learning is used to automatically create or refine weights and to create clauses in an MLN. Weights can be learned generatively or discriminatively; clauses are learned using Inductive Logic Programming (ILP) [9]. The learning process makes repeated use of the inference phase, using one of the methods described below. However, it is common of many applications to use only the inference phase with an already configured knowledge base (KB) and a number of facts in relational tables, as is the case of the applications we use in our experiments in Sect. 4. Learning is not considered any further in the paper after this subsection.

In *generative* weight learning, the idea is to maximize the likelihood (a function of the parameters of our statistical model) of our training evidence following the closed-world assumption [12]. i.e.: all ground atoms not in the database are false. However, computing the likelihood requires all true groundings of each clause, a difficult task even for a single clause [34]. MLNs use pseudo-log likelihood instead [6], which consists of a logarithmic approximation of the likelihood. Combined with a good optimizer like L-BFGS [7], pseudo-log likelihood can create weights for domains with millions of groundings.

Discriminative weight learning is used to predict query atoms given that we know the value of other atoms. This is achieved by maximizing the conditional log-likelihood (CLL: a constrained version of the log-likelihood), instead of the pseudo-log likelihood [37]. Maximizing CLL can be performed by optimizer algorithms like Voted Perceptron, Diagonal Newton, Scaled Conjugate Gradient, among others [11].

Clauses can be learned using ILP algorithms. The most important difference is the use of an evaluation function based on pseudo-likelihood, rather than accuracy or coverage. These modified methods include top-down structure learning [18] (TDSL) and bottom-up structure learning [24] (BUSL). On real-world application against famous ILP systems like CLAUDIEN [10], FOIL [29] and Aleph [39], both TDLS and BUSL find better MLN clauses.

2.4   Tuffy

Tuffy [26] is an MLN system that employs a bottom-up approach to grounding that allows for a more efficient procedure, in contrast to the top-down approach used by other systems. It also performs an efficient local search using an RDBMS. Inference is performed using the MaxWalkSat algorithm mentioned in Sect. 2.1. Tuffy can also perform parameter learning, but it does not implement structure learning (creating new clauses). In order to speedup execution, it partitions the MRF formed by the grounded clauses so as to perform random walks in parallel for each partition.

2.5   Evaluation of Datalog Programs

Datalog programs can be evaluated through a top-down approach or a bottom-up approach. The top-down approach (used by Prolog) starts with the goal that is reduced to subgoals, or simpler problems, until a trivial problem is reached. It is tuple-oriented: each tuple is processed through the goal and subgoals using all relevant facts. Because evaluating each goal can give rise to very different computations, the top-down approach is not easily adapted to GPUs *bulk* parallelism — more on this below and in Sect. 2.6.

The bottom-up approach first applies the rules to the given facts, thereby deriving new facts, and repeats this process with the new facts until no more facts are derived. The query is considered only at the end, to select the facts matching the query. Based on relational operations (as described shortly), this approach is suitable for GPUs because such operations are set-oriented and relatively simple overall. Also, rules can be evaluated in any order. This approach can be improved using the magic sets transformation [5] or the subsumptive tabling

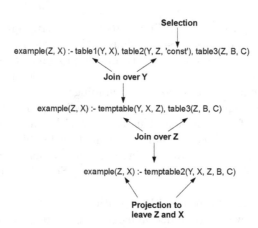

Fig. 1. Evaluation of a Datalog rule based on relational algebra operations.

transformation [40], through which the set of facts that can be inferred tends to contain only facts that would be inferred during a top-down evaluation.

Bottom-up evaluation of Datalog rules can be implemented with the relational algebra operators *selection*, *join* and *projection*, as outlined in Fig. 1. *Selections* are made when constants appear in the body of a rule. Next, a *join* is made between two or more subgoals in the body of a rule using the variables as reference. The result of a join can be seen as a temporary subgoal (or table) that has to be joined in turn to the rest of the subgoals in the body. Finally, a *projection* is made over the variables in the head of the rule.

We use fixed-point evaluation to compute recursive rules [41]. The basic idea is to iterate through the rules in order to derive new facts, and using these new facts to derive even more new facts until no new facts are derived.

2.6 GPU Architecture and Programming

GPUs are high-performance many-core processors capable of very high computation and data throughput [2]. They are used in a wide variety of applications [3]: games, data mining, bioinformatics, chemistry, finance, imaging, weather forecast, etc. Applications are usually accelerated by at least one order of magnitude, but accelerations of 10 times or more are common.

GPUs are akin to single-instruction-multiple-data (SIMD) machines: they consist of many processing elements that run the *same program* but on distinct data items. This program, referred to as the *kernel*, can be quite complex including control statements such as *if* and *while* statements. A kernel is executed by groups of threads called *warps* [1]. These warps execute one common instruction at a time, so all threads of a warp must have the same execution path in order to obtain maximum efficiency. If some threads diverge, the warp serially executes each branch path, disabling threads not on that path, until all paths complete and the threads converge to the same execution path. Hence, if a kernel has to compare strings, processing elements that compare longer strings will take longer and other processing elements that compare shorter strings will have to wait.

GPU memory is organized hierarchically. Each (GPU) thread has its own *per-thread local* memory. Threads are grouped into *blocks*, each block having a memory *shared* by all threads in the block. Finally, thread blocks are grouped into a single *grid* to execute a kernel — different grids can be used to run different kernels. All grids share the *global memory*.

3   Our GPU-Based Markov Logic Platform

Our platform TuGPU was designed to accelerate the grounding step, as this is often the most time consuming. Its main components are: the Tuffy Markov logic system [26], the YAP Prolog system [35] and GPU-Datalog [22]. The latter evaluates Datalog programs with a bottom-up approach using GPU kernels that implement the relational algebra operations *selection, join* and *projection*. For GPU-Datalog to be able to run Markov logic networks, its original version was extended with: management of stratified negation; improved processing of built-in comparison predicates; processing of disjunctions, in addition to conjunctions (to simplify specifying SQL queries as Datalog queries and to improve their processing); and an interface to communicate directly with PostgreSQL.

Figure 2 shows the interaction between the main modules of our platform in running a Markov logic network. Tuffy is called first, receiving three input files: (i) the evidence (facts) file; (ii) the MLN file; and the queries. Tuffy starts by creating a temporary database in PostgreSQL to store the evidence data and partial results (left side of Fig. 2). It then parses the program and query files in order to determine predicates and to create a (relational) table for each predicate found. Tables are then loaded with the evidence data.

The original Tuffy would then start the grounding phase. In TuGPU, this phase is performed by GPU-Datalog (center of Fig. 2), but, as Tuffy uses conjunctions to

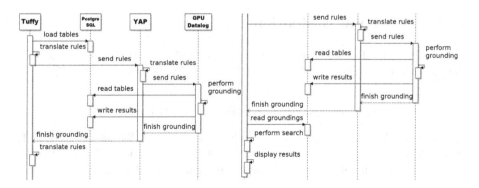

Fig. 2. TuGPU-Datalog modules running a Markov logic network. The left part corresponds to the active atoms grounding, while the right corresponds to the active clauses grounding.

specify a program, we first translate it to Datalog disjunctions. Then the Datalog program is sent to YAP, using a Java-Prolog interface, to compile it into a numerical representation (NR) where each unique string is assigned a unique integer id. YAP then sends the program's NR to GPU-Datalog to process the grounding. By using an NR, our GPU kernels show relatively short and *constant* processing time because all tuples in a table, being managed as sets of integers, can be processed in the same amount of time. Tuffy also uses an NR for evidence loaded in the database; this simplified extending it with GPU processing. Weights are not used in this phase, since the search for the most probable world will be performed by the host after the grounding is done.

To speed-up the grounding, Tuffy and TuGPU use the Knowledge-Based Model Construction [26] (KBMC) algorithm to determine those atoms and clauses that are relevant to the query. Then, GPU-Datalog reads the evidence from the database and performs the first step (of two) of the grounding process: computing the closure of the *active atoms* (i.e., those atoms whose truth value might change from true to false or vice versa, during search). The second step determines the *active clauses*, clauses that can be violated (i.e., their truth value becomes false) by flipping zero or more active atoms. For this step, TuGPU translates the program rules from the SQL that Tuffy generates into Datalog, and then YAP translates it into the NR used by GPU-Datalog.

When GPU-Datalog finishes each grounding step, it writes the found *active atoms* or *clauses* to the database. At the end of both grounding steps, Tuffy searches for the most likely world of the MLN. The search begins by using the ground active atoms and clauses to construct the MRF and then partition it into components. Each component has a subset of the active atoms and clauses, so that if an atom is flipped, it affects only those clauses found in the component.

The partitioned MRF is processed in parallel by the CPU with one thread per core and one component per thread, using the MaxWalkSAT algorithm mentioned in Sect. 2.1. The algorithm stops after a certain number of iterations or after an error threshold is reached. Finally, the results are displayed by TuGPU.

4   Experimental Evaluation

This section describes our experimental evaluation of the performance of TuGPU compared to that of the systems Alchemy [11], Tuffy [26] and RockIt [27].

4.1   Applications and Hardware-Software Platform

We used the following applications available with the Tuffy package. Table 1 shows some of their characteristics. For two of them (ER and RC), more tuples were randomly generated to test the systems with bigger data (right column).

- *Entity Resolution (ER)*: a simple, recursive MLN to determine if a person has cancer based on who his/her friends are and their smoking habits (this is an example from [31]).

Table 1. Applications characteristics.

Application	Inference rules	Evidence relations	Tuples in relations	
			Original	Random
ER	3	3	8	(310,000)
RC	15	4	82,684	(441,074)
IE	1024	18	255,532	(na)

- *Relational Classification (RC)*: classifies papers into 10 categories based on authorship and on the categories of other papers it references (Cora dataset [23]).
- *Information Extraction (IE)*: given a set of Citeseer citations, divided in tokens, rules with constants are used to extract structured records.

The original number of tuples of ER is only 8. We created another data set with a larger number of tuples, $310,000$, with randomly generated data: creating a fixed number of people, assigning a small random number of friends to each person, and labelling a fixed number of people as smokers.

For RC we also created a larger, randomly generated data set with $441,074$ tuples. We used a fixed number of papers and authors, and the same categories found in the original data: each author has a small random number of written papers, each paper is referred to by a small random number of other papers, and a small fixed number of papers are already labeled as belonging to a particular category.

We ran our experiments in the following hardware-software platform. **Host**: an AMD Opteron 6344, 12 cores CPU, with 64 GB DRAM. **GPU**: a Tesla K40c, 2880 CUDA Cores, with 12 GB GDDR5 memory and CUDA Capability 3.5. **Software**: CentOS 7, PostgreSQL 9.5 and CUDA Toolkit 7.0.

4.2 Results

Figure 3 shows the performance of the systems using the original datasets available in the Tuffy package and on our extended versions of these datasets. The left side shows our system to be the fastest in 2 out of the 3 original datasets, but only by a few seconds relative to standard Tuffy. Alchemy was the fastest in ER because the dataset is small and does not incur overhead setting up a database. We were unable to execute IE in RockIt, hence the empty space in the graph. Figure 3 (right) shows the performance of the systems with the extended datasets. For ER, our system was 15 times faster than RockIt and 77 times faster than Alchemy. Tuffy did not finish the grounding after more than 3 h. RockIt was 2.5 times faster than our system for RC. Both Tuffy and Alchemy did not finish after more than 5 h.

We performed a detailed analysis to determine why our system performed so well in ER and so poorly in RC. For ER, our random data, combined with

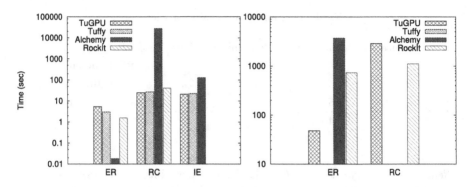

Fig. 3. Performance of the systems with original (left) and random (right) datasets. Note that the graphs are in log. scale.

its recursive clauses, generates many more recursive steps, 24 vs 2 in the original data. Each recursive step creates new tuples that need to be evaluated again. In our system, approximately 1,000,000 new tuples were generated in each iteration, most of them to be later discarded by our duplicate elimination kernel. Since our system was designed around these recursive applications, grounding was finished rather quickly while other systems struggled with costly joins that do not capitalize on parallel processing.

In RC, the number of recursive steps was 2 for both normal and random datasets. We hence analyzed the execution times of each part of our system. Using our random data, both atom and clause grounding take about 2 min to complete, loading data and other tasks take 30 s, but the search phase takes an astounding 43 min. In contrast, the times for ER are about 8 s for both groundings, 21 s for data loading and other tasks, and 16 s for the search.

Also in the search phase, ER, despite generating many more intermediate tuples during grounding, uses only 252,249 active clauses, while RC uses 5,586,900. Furthermore, when partitioning the resulting MRF, we get a single component with approximately 4,000,000 active clauses. Since each component is assigned to a thread (and one thread to each CPU-core), smaller components finish quickly and we are left with a very large component being processed by a single core, thus dominating the execution time. In contrast, RockIt creates a large optimization problem but its parallel resolution has a much better balanced workload.

Overall these results are promising since they mean that the benefit of performing the grounding phase on the GPU outweighs the overhead associated with the database and GPU I/O, even for rather small datasets.

5   Related Work

The wide adoption of Markov logic for various types of applications has fostered the development of various systems and research on improvements. Alchemy was the first Markov logic system implementation [11]. It is one of the most complete systems, including various algorithms for inference following a top-down approach, various techniques for learning weights and structure and more. The original Alchemy always performs inference by first grounding the program and then using approximated methods either based in MCMC (Markov Chain Monte Carlo), such as MC-SAT and Gibbs sampling, or variants of belief propagation. Alchemy supports both total probability and most likely explanation queries, and also provides a large number of learning algorithms. However, Alchemy does not cope well with large real-world applications.

Tuffy was developed by Feng Niu *et al.* [26]. It relies on PostgreSQL relational database management system (RDBMS) to perform inference. Tuffy follows a bottom-up approach to solve the grounding step. This allows the grounding to be expressed as SQL queries which, combined with query optimization by the RDBMS, allows Tuffy to complete the grounding faster than Alchemy.

Several other systems are available. theBeast, developed by Riedel [32], uses Cutting Planes Inference (CPI) optimization, which instantiates and solves small parts of a complex MLN network. This takes advantage of the observation that inference can be seen as either a MAX-SAT problem or as an integer linear programming problem (i.e. a mathematical optimization problem where the variables are restricted to integers). While theBeast is faster than Alchemy for some problems, it lacks many of Alchemy's features such as structure learning and MPE (Most Probable Explanation) inference [21].

RockIt is a recent system by Noessner *et al.* [27]. It treats the inference problem as an integer linear programming problem and includes a new technique called cutting plane aggregation (CPA) which, coupled with shared-memory multi-core parallelism during most of the inference, allows RockIt to outperform all other systems.

Beedkar *et al.* implemented fully parallel inference for MLNs [4]. Their system parallelizes grounding by considering each clause as a set of joins and partitioning them according to a single *join graph*. The search step of inference is also parallelized using importance sampling together with MCMC [20]. Since the MLN is partitioned during grounding, no further partitioning is required before searching. This approach is more efficient than Tuffy's since the partition is performed over a smaller, data independent graph. Experimental evaluation shows that this is faster and produces similar results when compared with Tuffy.

Other works speedup inference and learning with MLNs. Shavlik and Natarajan [36] propose ways of speeding up inference by using a preprocessing algorithm that can substantially reduce the effective size of MLNs by rapidly counting how often the evidence satisfies each formula, regardless of the truth values of the query atoms. Mihalkova and Mooney [24] and Davis and Domingos [8] have proposed bottom-up methods that can improve structure learning time and accuracy over existing top-down approaches. Mihalkova and Richardson [25] proposes to cluster query atoms and then perform full inference for only one representative from each cluster.

Our system is the first one to run Markov logic networks using GPUs. Since Datalog and MLNs share an equivalent syntax, a modified version of our

GPU-Datalog engine was used. Like Tuffy, our system uses a bottom-up approach based on relational operators to process one of the most time consuming parts of the inference step, but in a GPU.

Similar to our work on GPU-Datalog (described in Sect. 3), Wu *et al.* created Red Fox [44], a system that parallelizes relational algebra and other operations in the GPU, in order to solve programs based on a variant of Datalog called LogiQL. Comparison with GPU-Datalog using the famous TCP-H queries can be found in [22]. Other similar systems that execute SQL queries in parallel using the GPU include [14, 45].

6   Conclusions

We have presented a system that accelerates the grounding step in MLNs by combining Tuffy with our GPU-Datalog engine. Its performance is on par or better than other well-known MLN systems. Our results show that the benefit of performing the grounding phase on the GPU outweighs the overhead of using a database and of GPU I/O, even for rather small datasets. Our system can be greatly improved by also performing the search step of the inference phase in the GPU. This would require the parallelization of a SAT solver. There are several available in the literature and we expect to benefit from extensive work in parallelization of SAT and ILP solvers.

Our GPU-Datalog system could benefit from data partitioning algorithms. This would allow tables bigger than the amount of GPU memory available to be processed.

Since GPU-Datalog has been successfully used to improve ILP [22], we believe that clause learning in MLNs could also be improved by our system. We also plan to research the parallelization of generative and discriminative weight learning.

Acknowledgments. This work is partially financed by the ERDF - European Regional Development Fund through the Operational Programme for Competitiveness and Internationalisation - COMPETE 2020 Programme within project <<POCI-01-0145-FEDER-006961>>, and by National Funds through the FCT - Fundação para a Ciência e a Tecnologia (Portuguese Foundation for Science and Technology) as part of project UID/EEA/50014/2013. Also, Martínez-Angeles gratefully acknowledges grants from CONACYT and Cinvestav-IPN. Finally, we gratefully acknowledge the major contribution of the referees, whose recommendations have led us to significantly improve the paper.

References

1. Cuda C Programming Guide. http://docs.nvidia.com/cuda/cuda-c-programming-guide/index.html
2. General-Purpose Computation on Graphics Hardware, March 2015. http://gpgpu.org/
3. GPU Applications, March 2015. http://www.nvidia.com/object/gpu-applications-domain.html

4. Beedkar, K., Del Corro, L., Gemulla, R.: Fully parallel inference in markov logic networks. In: 15th GI-Symposium Database Systems for Business, Technology and Web, BTW 2013. Bonner Kllen, Magdeburg (2013)
5. Beeri, C., Ramakrishnan, R.: On the power of magic. J. Log. Program. **10**(3–4), 255–299 (1991)
6. Besag, J.: Statistical analysis of non-lattice data. J. R. Stat. Soc. Ser. D (Stat.) **24**(3), 179–195 (1975)
7. Byrd, R.H., Lu, P., Nocedal, J., Zhu, C.: A limited memory algorithm for bound constrained optimization. SIAM J. Sci. Comput. **16**(5), 1190–1208 (1995)
8. Davis, J., Domingos, P.: Bottom-up learning of markov network structure. In: Proceedings of the Twenty-Seventh International Conference on Machine Learning, pp. 271–280. ACM Press (2010)
9. De Raedt, L.: Logical and Relational Learning. Springer, Heidelberg (2008)
10. De Raedt, L., Dehaspe, L.: Clausal discovery. Mach. Learn. **26**(2–3), 99–146 (1997)
11. Domingos, P., Lowd, D.: Markov Logic: An Interface Layer for Artificial Intelligence, 1st edn. Morgan and Claypool Publishers, San Rafael (2009)
12. Genesereth, M.R., Nilsson, N.J.: Logical Foundations of Artificial Intelligence. Morgan Kaufmann Publishers Inc., San Francisco (1987)
13. Getoor, L., Taskar, B.: Introduction to Statistical Relational Learning. MIT Press, Cambridge (2007)
14. He, B., et al.: Relational query coprocessing on graphics processors. ACM Trans. Database Syst. (TODS) **34**(4), 21 (2009)
15. Huang, S.S., et al.: Datalog and emerging applications: an interactive tutorial. In: SIGMOD Conference, pp. 1213–1216 (2011)
16. Kautz, H., Selman, B., Jiang, Y.: A general stochastic approach to solving problems with hard and soft constraints. In: The Satisfiability Problem: Theory and Applications, pp. 573–586. American Mathematical Society (1996)
17. Kindermann, R., Snell, J.L.: Markov Random Fields and their Applications, 1st edn. American Mathematical Society, Providence (1980)
18. Kok, S., Domingos, P.: Learning the structure of markov logic networks. In: Proceedings of the 22Nd International Conference on Machine Learning, ICML 2005, pp. 441–448. ACM, New York (2005)
19. Kok, S., Domingos, P.: Extracting semantic networks from text via relational clustering. In: Daelemans, W., Goethals, B., Morik, K. (eds.) ECML PKDD 2008, Part I. LNCS (LNAI), vol. 5211, pp. 624–639. Springer, Heidelberg (2008)
20. Koller, D., Friedman, N.: Probabilistic Graphical Models. MIT Press, Cambridge (2009)
21. Kwisthout, J.: Most probable explanations in bayesian networks: complexity and tractability. Int. J. Approx. Reason. **52**(9), 1452–1469 (2011)
22. Martínez-Angeles, C.A., Wu, H., Dutra, I., Santos-Costa, V., Buenabad-Chávez, J.: Relational learning with GPUs: accelerating rule coverage. Intl. J. Parallel Programm. **44**(3), 663–685 (2016). doi:10.1007/s10766-015-0364-7. http://dx.doi.org/10.1007/s10766-015-0364-7
23. McCallum, A., Nigam, K., Rennie, J., Seymore, K.: Automating the construction of internet portals with machine learning. Inf. Retrieval **3**(2), 127–163 (2000)
24. Mihalkova, L., Mooney, R.J.: Bottom-up learning of markov logic network structure. In: Proceedings of the Twenty-Fourth International Conference on Machine Learning, pp. 625–632. ACM Press (2007)
25. Mihalkova, L., Richardson, M.: Speeding up inference in statistical relational learning by clustering similar query literals. In: Raedt, L. (ed.) ILP 2009. LNCS, vol. 5989, pp. 110–122. Springer, Heidelberg (2010)

26. Niu, F., Ré, C., Doan, A., Shavlik, J.: Tuffy: scaling up statistical inference in markov logic networks using an rdbms. Proc. VLDB Endow. **4**(6), 373–384 (2011)
27. Noessner, J., Niepert, M., Stuckenschmidt, H.: Rockit: exploiting parallelism and symmetry for MAP inference in statistical relational models. CoRR, abs/1304.4379 (2013)
28. Poon, H., Domingos, P., Sumner, M.: A general method for reducing the complexity of relational inference and its application to mcmc. In: Proceedings of the 23rd National Conference on Artificial Intelligence, AAAI 2008, vol. 2, pp. 1075–1080. AAAI Press (2008)
29. Quinlan, J.R.: Learning logical definitions from relations. Mach. Learn. **5**(3), 239–266 (1990)
30. De Raedt, L., Kersting, K.: Probabilistic inductive logic programming. In: Raedt, L., Frasconi, P., Kersting, K., Muggleton, S.H. (eds.) Probabilistic Inductive Logic Programming. LNCS (LNAI), vol. 4911, pp. 1–27. Springer, Heidelberg (2008)
31. Richardson, M., Domingos, P.: Markov logic networks. Mach. Learn. **62**(1–2), 107–136 (2006)
32. Riedel, S.: Improving the accuracy and efficiency of map inference for markov logic. In: Proceedings of the 24th Annual Conference on Uncertainty in AI, UAI 2008, pp. 468–475 (2008)
33. Riedel, S., Meza-Ruiz, I.: Collective semantic role labelling with markov logic. In: Proceedings of the Twelfth Conference on Computational Natural Language Learning, CoNLL 2008, pp. 193–197. Association for Computational Linguistics, Stroudsburg (2008)
34. Roth, D.: On the hardness of approximate reasoning. Artif. Intell. **82**(1–2), 273–302 (1996)
35. Santos-Costa, V., et al.: The YAP Prolog system. TPLP **12**(1–2), 5–34 (2012)
36. Shavlik, J., Natarajan, S.: Speeding up inference in markov logic networks by preprocessing to reduce the size of the resulting grounded network. In: Proceedings of the 21st International Jont Conference on Artifical Intelligence, IJCAI 2009, pp. 1951–1956. Morgan Kaufmann Publishers Inc., San Francisco (2009)
37. Singla, P., Domingos, P.: Discriminative training of markov logic networks. In: Proceedings of the 20th National Conference on Artificial Intelligence, AAAI 2005, vol. 2, pp. 868–873. AAAI Press (2005)
38. Singla, P., Domingos, P.: Lifted first-order belief propagation. In: Proceedings of the 23rd National Conference on Artificial Intelligence, AAAI 2008, vol. 2, pp. 1094–1099. AAAI Press (2008)
39. Srinivasan, A.: The Aleph Manual (2001)
40. Tekle, K.T., Liu, Y.A.: More efficient datalog queries: subsumptive tabling beats magic sets. In: SIGMOD Conference, pp. 661–672 (2011)
41. Ullman, J.: Principles of Database and Knowledge-Base Systems, vol. I. Computer Science Press, Rockville (1988)
42. Ullman, J.: Principles of Database and Knowledge-Base Systems, vol. II. Computer Science Press, Rockville (1989)
43. Wu, F., Weld, D.S.: Automatically refining the wikipedia infobox ontology. In: Proceedings of the 17th International Conference on World Wide Web, WWW 2008, pp. 635–644. ACM, New York (2008)
44. Wu, H., Diamos, G., Sheard, T., Aref, M., Baxter, S., Garland, M., Yalamanchili, S.: Red fox: an execution environment for relational query processing on gpus. In: International Symposium on Code Generation and Optimization (CGO) (2014)
45. Yuan, Y., Lee, R., Zhang, X.: The yin and yang of processing data warehousing queries on gpu devices. Proc. VLDB Endow. **6**(10), 817–828 (2013)

Using ILP to Identify Pathway Activation Patterns in Systems Biology

Samuel R. Neaves[1]([✉]), Louise A.C. Millard[2,3], and Sophia Tsoka[1]

[1] Department of Informatics, King's College London, Strand, London, UK
{samuel.neaves,sophia.tsoka}@kcl.ac.uk
[2] MRC Integrative Epidemiology Unit (IEU) at the University of Bristol,
University of Bristol, Bristol, UK
louise.millard@bristol.ac.uk
[3] Intelligent Systems Laboratory, Department of Computer Science,
University of Bristol, Bristol, UK

Abstract. We show a logical aggregation method that, combined with propositionalization methods, can construct novel structured biological features from gene expression data. We do this to gain understanding of pathway mechanisms, for instance, those associated with a particular disease. We illustrate this method on the task of distinguishing between two types of lung cancer; Squamous Cell Carcinoma (SCC) and Adenocarcinoma (AC). We identify pathway activation patterns in pathways previously implicated in the development of cancers. Our method identified a model with comparable predictive performance to the winning algorithm of a recent challenge, while providing biologically relevant explanations that may be useful to a biologist.

Keywords: Biological pathways · Warmr · TreeLiker · Reactome · Barcode · Logical aggregation

1 Introduction and Background

In the field of Systems Biology researchers are often interested in identifying perturbations within a biological system that are different across experimental conditions. Biological systems consist of complex relationships between a number of different types of entities, of which much is already known [1]. An Inductive Logic Programming (ILP) approach may therefore be effective for this task, as it can represent the relationships of such a system as background knowledge, and use this knowledge to learn potential reasons for differences in a particular condition. We demonstrate a propositionalization-based ILP approach, and apply this to the example of identifying differences in perturbations between two types of lung cancer; Squamous Cell Carcinoma (SCC) and Adenocarcinoma (AC).

A recent large competition run by the SBV Improver organisation, called the Diagnostic Signature Challenge, tasked competitors with finding a highly predictive model distinguishing between these two lung cancer types [2]. The challenge

© Springer International Publishing Switzerland 2016
K. Inoue et al. (Eds.): ILP 2015, LNAI 9575, pp. 137–151, 2016.
DOI: 10.1007/978-3-319-40566-7_10

was motivated by the many studies that have also worked on similar tasks, with the aim to find a model with the best predictive performance. The winning method from this competition is a pipeline that exemplifies the classification approaches used for this task [3].

The typical pipeline has three distinct stages. The first stage uses technology such as microarrays or RNAseq, to measure gene expressions across the genome in a number of samples from each of the experimental conditions. The second stage identifies a subset of genes whose expression values differ across conditions. This stage is commonly achieved by performing differential expression analysis and ranking genes by a statistic such as fold change values [4]. A statistical test is then used to identify the set of genes to take forward to stage 3. Alternatively for stage 2, researchers may train a model using machine learning to classify samples into experimental conditions, often using an attribute-value representation where the features are a vector of gene expression values (as performed by the winning SBV Improver model). This approach has the advantage that the constructed model may have found dependencies between genes that would not have been identified otherwise. Researchers use the 'top' features from the model to identify the set of genes to take forward to stage 3.

In stage 3 researchers look for connections between these genes by, for example, performing a Gene Set Enrichment Analysis (GSEA) [5]. Here, the set of genes are compared with predefined sets of genes, that each indicate a known relation. For example, a gene set may have a related function, exist in the same location in the cell, or take part in the same pathway.

To bring background knowledge of relations into the model building process, past ILP research integrated stage 2 (finding differentially expressed genes) and stage 3 (GSEA), into a single step [6]. This was achieved using Relational Subgroup Discovery, which has the advantage of being able to construct novel sets by sharing variables across predicates that define the sets. For example, a set could be defined as the genes annotated with two Gene Ontology terms.

Other ways researchers have tried to integrate the use of known relations includes adapting the classification approach of stage 2. New features are built by aggregating across a predefined set of genes. For example, an aggregation may calculate the average expression value for a pathway [7].

A major limitation of current classification approaches is that the models are constructed from either genes or crude aggregates of sets of genes, and so ignore the detailed relations between entities in a pathway. In order to incorporate more complex relations an appropriate network representation is needed, such that biological relations are adequately represented. For example, a simple directed network of genes and proteins does not represent all the complexities of biochemical pathways, such as the dependencies of biochemical reactions. To do this bipartite graphs or hypergraphs can be used [8].

One way to incorporate more complex relations is by creating topologically defined sets, where a property of the network is used to group nodes into related sets. One method to generate these sets is Community Detection [9]. However, this approach can create crude clusters of genes, that do not account for important biological concepts. Biologists may be interested in complex biological interactions rather than just sets of genes.

Network motif and frequent subgraph mining are methods that can look for structured patterns in biological networks [10]. However, in these approaches the patterns are often described in a language which is not as expressive as first order logic. This means they are unable to find patterns with uninstantiated variables, or with relational concepts such as paths or loops.

To our knowledge only one previous work has used ILP for this task [11]. Here the authors propose identifying features consisting of the longest possible chain of nodes in which non-zero node activation implies a certain (non-zero) activation in its successors, which they call a Fully Coupled Flux. Their work is preliminary, with limited evaluation of the performance of this method.

The aim of this paper is to illustrate how we can identify pathway activation patterns, that differ between biological samples of different classes. A pathway activation pattern is a pattern of active reactions on a pathway. Our novel approach uses known relations between entities in a pathway, and important biological concepts as background knowledge. These patterns may give a biologist different information than models built from simple gene features. We seek to build models that are of comparative predictive performance to those of previous work, while also providing potentially useful explanations.

In this work we take a propositionalization-based ILP approach, where we represent the biological systems as a Prolog knowledge base (composed of first-order rules and facts), and then reduce this to an attribute-value representation (a set of propositions), before using standard machine learning algorithms on this data. We begin with an overview of propositionalization, and a discussion of why it is appropriate for this task.

2 Overview of Propositionalization

Propositionalization is a method that transforms data represented in first-order logic to a set of propositions, i.e. a single table representation where each example is represented by a fixed length vector. This is called a reduction. It is possible to make a proper reduction of the representation using Goedel numbering or well-ordering arguments [12]. However, these will have limited practical value as useful structure can be lost or encoded inefficiently, leading to poor inductive ability. Heuristic-based propositionalization methods allow specification of a language bias and a heuristic, in order to search for a subset of potential features which are useful for the learning task.

We have four reasons for adopting a propositionalization-based approach, rather than directly applying an ILP learner. First, separating the feature construction from the model construction means that we have an interesting output in the middle of the process, which we would lose if they were coupled together. For example, the features constructed can represent useful domain knowledge in their own right, as they can describe subgroups of the data which have a different class distribution, or frequent item sets or queries on the data.

Second, propositionalization can be seen as a limited form of predicate invention, where the predicate refers to a property of an individual, or relationships amongst properties of the individual. This means that, when building a model, the features may correspond to complex relationships between the original properties of an individual. In our case they correspond to potentially interesting pathway activation patterns. Hence, we can understand predictions in terms of these higher order concepts, which may give important insights to a biologist.

Third, propositionalization can impose an individual-centred learning approach [12,13]. This limits predicates to only refer to relationships between properties of an individual – we cannot have a predicate which relates individuals. This strong inductive bias is appropriate for our case, as we do not wish to consider relationships between the individuals. The fourth reason is that we can perform many other learning tasks on the transformed data, with the vast array of algorithms available for attribute-values datasets.

In this work we use query-based propositionalization methods, and now describe some key algorithms. A review of some publicly available propositionalization methods was recently performed by Lavrač et al. [14]. These include Linus, RSD, TreeLiker (HiFi and RelF algorithms), RELAGGS, Stochastic Propositionalization, and Wordification, alongside the more general ILP toolkit, Aleph. Other methods that were not mentioned in that review include Warmr [15], Cardinalisation [16], ACORA [17] and CILP++ [18]. There has also been work on creating propositionalization methods especially for linked open data, both in an automatic way [19], and in a way where manual SPARQL queries are made [20]. The methods in these papers are not appropriate for our work because our data is not entirely made up of linked open data, and we wish to include background rules encoding additional biological knowledge. It is also worth noting that certain kernel methods can be thought of as propositionalization [12].

Wordification treats relational data as documents and constructs word-like features. These are not be appropriate for our task, as they do not correspond to the kind of patterns we are looking for, i.e. features with uninstantiated variables. Stochastic propositionalization performs a randomised evolutionary search for features. This approach may be interesting to consider for future work. CILP++ is a method for fast bottom-clause construction, defined as the most specific clause that covers each example. This method is primarily designed to facilitate the learning of neural networks, and has been reported to perform no better than RSD when used with a rule-based model [18].

ACORA, Cardinalisation and RELAGGS are database inspired methods of propositionalization. They are primarily designed to perform aggregation across a secondary table, with respect to a primary table. ACORA is designed to create aggregate operators for categorical data, whereas RELAGGS performs standard aggregation functions (summation, maximum, average etc.) suitable for numeric data. Cardinalisation is designed to use complex aggregates, where conditions are added to an aggregation. In our work we manually design an aggregation method, described in Sect. 3.2. These aggregation systems are not appropriate for graph-based datasets, because representing the graph as two tables (denoting edges and nodes) and aggregating on paths through the graph would require many self joins on the edge table. Relational databases are not optimised for this task, such that the resulting queries would be inelegant and inefficient.

The propositionalization methods we use in this work are TreeLiker and Warmr. TreeLiker is a tool that provides a number of algorithms for propositionalization including RelF [21]. RelF searches for relevant features in a block-wise manner, and this means that irrelevant and irreducible features can be discarded during the search. The algorithms in TreeLiker are limited to finding tree-like features where there are no cycles. RelF has been shown to scale much better than previous systems such as RSD, and can learn features with tens of literals. This is important for specifying non-trival pathway activation patterns.

Warmr is a first-order equivalent of frequent item-set mining, where a level-wise search of frequent queries in the knowledge base is performed. Warmr is used as a propositionalization tool by searching for frequent queries in each class. In Warmr it is possible to specify the language bias using conjunctions of literals, rather than just individual literals, and to put constraints on which literals are added. This allows strong control of the set of possible hypotheses that can be considered. Finally, unlike TreeLiker, Warmr can use background knowledge, defined as facts and rules.

3 Methods

An overview of the process we take is shown in Fig. 1. First, we extract the reaction graph for each pathway, from Reactome. Second, we infer the instantiated reaction graphs for each instance in the dataset. Third, we identify pathway activation patterns using propositionalization, and then build classification models to predict the lung cancer types. Lastly, we evaluate our models using a hold-out dataset. We begin with a description of the datasets we use in this work.

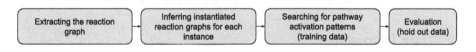

Fig. 1. Method overview

3.1 Raw Data

Our approach uses two sources of data: (1) a dataset from Gene Expression Omnibus (GEO) [22] as the set of examples (gene expression values of a set of individuals), and (2) information about biological systems from Reactome.

GEO Data. We use a two class lung cancer dataset obtained from GEO, which was previously used in the SBV Improver challenge [2]. This dataset is made up from the following datasets: GSE2109, GSE10245, GSE18842 and GSE29013 (n=174), used as training data, and GSE43580 (n=150), used as hold-out data. We used the examples where the participants were labelled as having either SCC or AC lung cancer. This is the same data organisation as that used in SBV

Improver challenge, to allow us to compare our results with the top performing method from this challenge.

This data contains gene expression measurements from across the genome measured by Affymetrix chips. Each example is a vector of 54,614 real numbers. Each value denotes the amount of expression of mRNA of a gene. There is a uniform class distribution of examples, in both the training and holdout dataset.

Reactome-Background Knowledge. We use the Reactome database to provide the background knowledge, describing biological pathways in humans. Reactome [1] is a collection of manually curated peer reviewed pathways. Reactome is made available as an RDF file, which allows for simple parsing using SWI-Prolog's semantic web libraries, and contains 1,351,811 triples. Reactome uses the bipartite network representation of entities and reactions. Entity types include nucleic acids, proteins, protein complexes, protein sets and small molecules. Protein complexes and protein sets can themselves comprise of other complexes or sets. In addition, a reaction may be controlled (activated or inhibited) by particular entities. A reaction is a chemical event, where input entities (known as substrates), facilitated by enzymes, form other entities (known as products).

Figure 2a shows a simple illustration of a Reactome pathway. P nodes denote proteins or protein complexes, R nodes denote reactions, and C nodes denote catalysts. A black arrow illustrates that a protein is an input or output of a reaction. A green arrow illustrates that an entity is an activating control for a reaction. A red arrow illustrates that an entity is an inhibitory control for a reaction. Reaction $R1$ has 3 protein substrates and 3 protein products, and is controlled by catalyst C. Reactions $R3$ and $R4$ both have one protein substrate and one protein product. $R3$ is inhibited by $P2$, such that if $P2$ is present then reaction $R3$ will not occur. $R4$ is activated by $P3$, such that $P3$ is required for reaction $R4$ to occur.

3.2   Data Processing

Extracting Reaction Graphs. We reduce the Reactome bipartite graph to a boolean network of reactions. This simplifies the graphs while still adequately encoding the relationships between entities. Previous work has shown that boolean networks are a useful representation of biological systems [23], and unlike gene and protein boolean networks ours encodes the dependencies between reactions.

The boolean networks we create are reaction-centric graphs, where nodes are reactions and directed edges are labelled either as 'activation', 'inhibition' or 'follows' corresponding to how reactions are connected. For example, Fig. 2b shows the reaction-centric graph, corresponding the Reactome graph shown in Fig. 2a. Reaction $R2$ follows R1, because in the Reactome graph $P1$ is an output of $R1$ and an input to $R2$. Reaction $R1$ inhibits $R3$, because $P2$ is an output of $R1$, and it is also an inhibitory control of $R3$. Reaction $R1$ activates reaction $R4$, because $P3$ is an output of $R1$, and an activating control of $R4$.

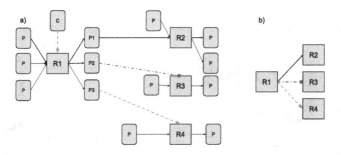

Fig. 2. Reaction Graph illustrations. There are three types of relationships between reactions: follows (black solid lines), activation (green dashed), and inhibition (red dash-dotted). (Color figure online)

Inferring Instantiated Reaction Graphs. Boolean networks [23] are a common abstraction in biological research, but these are normally applied at the gene or protein level not at the reaction level. In order to use a boolean network abstraction on a reaction network, we apply a logical aggregation method that aggregates measured probe values (from the GEO dataset) into reactions. This creates a binary value for each reaction, to create instantiated versions of the reaction-centric graph created in the previous step.

Before we can use this logical aggregation we first transform the original probe values into binary values, an estimated value denoting whether a gene is expressed or not. We do this using Barcode [24], a tool for converting the continuous probe values to binary variables, by applying previously learnt thresholds to microarray data. It is important to note that Barcode makes it possible to compare gene expressions, both within a sample, and between samples that are potentially measured by different arrays.

The logical aggregation process is illustrated in Fig. 3. This process takes the binary probe values as input, and uses the structure provided by the Reactome graph, and key biological concepts, to build reaction level features. As we have already described, each reaction has a set of inputs that are required for a particular reaction. We interpret each reaction input as a logical circuit with the following logical rules. The relationship between probes and proteins is treated as an OR gate (matched by Uniprot IDs), because multiple probes can encode for same protein. We are assuming that the measurement from a single probe indicates with high probability whether the protein product is present or not. The formation of a protein complex requires all of its constituent proteins and therefore is treated as an AND gate. A protein set is a set of molecules that are functionally equivalent such that only one is needed for a given reaction, and so this is treated as an OR gate. Inputs to a reaction are treated as an AND gate. A reaction is *on* if the inputs are *on*, any activating agents are *on*, and any inhibitory agents are *off*. We note that both protein sets and protein complexes can themselves comprise of arbitrarily nested complexes or sets.

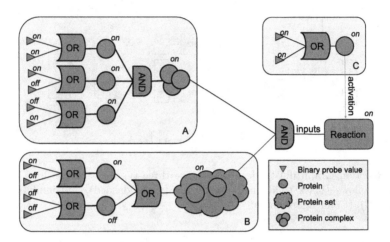

Fig. 3. Illustration of logical aggregation. Known biological mechanisms can be represented as *OR* or *AND* gates. The triangular nodes are binary probe values, created using barcode.

Figure 3 illustrates the logical aggregation rules of a single reaction. This reaction has two inputs and one activating control. The two inputs are a protein complex and a protein set, and the values of these are calculated using their own aggregation processes, labelled A and B. The aggregation in process A, starts with the binary probe values, and first infers the values of three proteins. The protein complex is then assigned a value of *on* because all proteins required for this complex are present (are all *on* themselves). The aggregation in process B starts by inferring the values of two proteins from the probe values. One protein is *on* and the other is *off*. The protein set is assigned the value *on* because only one protein in this set is required for this protein set to be *on*. There also exists an activating control for the reaction, a protein whose value is determined by a process labelled C in this figure. This protein is assigned the value *on*, because both probe values are *on*, when at least one is required. As all inputs are *on* and the activating control is also *on*, the reaction is assigned the value *on*.

3.3 Searching for Pathway Activation Patterns

In order to identify pathway activation patterns we first find pathways that are most likely to contain these patterns, using the training data. We then use three approaches to identify pathway activation patterns within the 'top' pathways, and evaluate the identified activation patterns using the hold-out data.

Identifying Predictive Pathways. To identify pathways we first run Tree-Liker on each pathway. This generates a set of attribute-value features for each instantiated pathway. We use TreeLiker with the RelF algorithm and the following language bias:

```
set(template,[reaction(-R1,#onoroff),link(+R1,-R2,!T1),
reaction(+R2,#onoroff),link(+R2,-R3,!T2),reaction(+R3,#onoroff),
link(!RA,-R4,!T3),link(+R4,!RB,!T4),link(+R1,-R2,#T1),
link(+R2,-R3,#T2),link(!RA,-R4,#T3),link(+R4,!RB,#T4)])
```

This language bias contains two types of literals; *reaction/2* and *link/3*. The second argument of the reaction literal is always constrained to be a constant depicting if a reaction is *on* or *off*. The *link* literal depicts the relationship between two reactions, where the third argument of the link literal can be a variable or a constant describing the type of relationship - either follows, activates or inhibits. For example, an identified pattern may contain the literal *link(r1,r2,follows)*, specifying that an output entity of reaction *r1* is an input to reaction *r2*.

We then test the performance of the features of each pathway, using 10 fold cross validation. We use the J48 decision tree algorithm (from Weka) because this builds a model that give explanations for the predictions. We calculate the average accuracy across folds, for each pathway, and rank the pathways from highest to lowest accuracy. We then use the top ranked pathways as input to three different methods, to identify predictive pathway activation patterns.

Method 1. This approach simply takes a pathway of interest, generates a single model using the J48 algorithm using the training data, and then evaluates this performance on the hold-out data. The decision tree can then be viewed to determine which activation patterns are predictive of lung cancer type. We demonstrate this approach with the top-ranked pathway.

Method 2: Warmr Approach. We illustrate using Warmr to generate pathway activation patterns, using one of our identified 'top' pathways.

We use Warmr with two particular concepts in the background knowledge. First, we use a predicate *longestlen/3*, that calculates the longest length of *on* reactions in an example, for the pathway on which Warmr is being run. The arguments are: (1) the beginning reaction of a path, (2) the end reaction of the path with longest length, and (3) the length of this path. This longest length concept corresponds to the fully coupled flux of a previous work [11].

Second, we use the predicates *inhibloop/1* and *actloop/1*, that depict inhibition and activation loops, where a path of *on* reactions form a loop and one of the edges is an inhibition or activation edge, respectively. Inhibition and activation loops are common biological regulatory mechanisms [25].

We then use the OneR (Weka) algorithm to identify the single best pathway activation pattern found by Warmr, and then evaluate this pattern on the hold-out data.

Method 3: Combined Approach. Our combined method takes advantage of the beneficial properties of the two algorithms, by using Warmr to extend the patterns identified by TreeLiker. This effectively switches the search strategy from the block-wise approach of TreeLiker, to the level-wise approach of Warmr. The reason for doing this is to identify any relations between reactions that exist between entities within the TreeLiker feature, that could not be identified in TreeLiker due to its restriction to tree structures. This results in long, cyclical features that neither TreeLiker nor Warmr would be able to find on their own.

While we could use the features generated by method 1, and extend these, in this section we also demonstrate the possibility of using our approach for generating descriptions of subgroups. We identify a subgroup with the CN2SD algorithm [26], using the training data. The activation patterns defining this subgroup are then extended using Warmr. The following code is an example language bias we use in Warmr:

```
rmode(1: (r(+S,-A,1),link(A,-\B,follows),link(B,-\C,_),r(S,C,0),
  r(S,B,0), link(B,-\D,_),r(S,D,1),link(A,-\E,_),r(S,E,1))).
rmode(1: link(+A,+B,#)).
```

The first *rmode* contains the feature that was previously identified using Tree-Liker. The second rmode uses the literal *link*, to allow Warmr to add new links to the TreeLiker feature. After extending the activation pattern using Warmr, we then evaluate this on the hold-out data.

4   Results

To reiterate, the aim of this work is to build explanatory models that help biologists understand the system perturbations associated with conditions, in this case lung cancer. Therefore, although we give the classification performance of our models in order to make the quantitative performance comparison, we additionally emphasis the form that the classification models take and how these are of interest to biologists. Table 1 shows the top 5 pathways found using our TreeLiker/J48 method, and the size of each reaction graph.

Table 1. Top 5 pathways identified. Mean accuracy across 10 folds of cross-validation on the training dataset.

Ranking	Pathway	Accuracy	Number of nodes (reactions)	Number of edges
1	Hexose uptake	78.74 %	18	25
2	Hyaluronan biosynthesis	77.59 %	18	25
4	Mitotic G1-G1/phases	76.74 %	51	59
4	Creatine metabolism	78.64 %	6	6
5	Cell Cycle	77.59 %	322	492

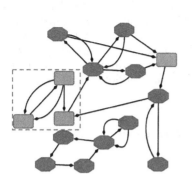 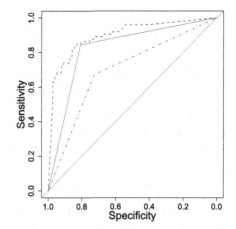

(a) Example of the hexose uptake pathway for a particular individual. Green
squares: *on* reactions, red octagons: *off*
reactions. Identified feature of three *on*
reactions shown in pink, dashed box.

(b) ROC curves comparing performance.
SBV Improver: blue, dashed; TreeLiker &
J48: green, solid; Warmr & OneR: red, dash-
dotted.

Fig. 4. Results (Color figure online)

4.1 Quantitative Evaluation and Comparison with SBV Improver Model

To provide a quantitative comparison of our models, we compare to the winning
classifier of the SBV Improver challenge. We use the area under the ROC curve
(AUC) metric, to evaluate the ranking performance of the models. We generate
confidence intervals for the AUC using a stratified bootstrapped approach (with
2000 bootstraps) [27]. We also use permutation testing to compare the performance of our models with a random model. We generate 2000 random rankings,
with the same class distribution as our data, and calculate the AUC for each of
these rankings. We then find the proportion of random rankings with an AUC
greater than that of our model. We refer to this as the permutation P value.

We select the top pathway identified in the training data – hexose uptake.
After retraining J48 on the whole training data, evaluation on the hold-out data
gives an AUC of 0.820 (95 % confidence interval (CI): 0.764–0.890). This model is
better than a random model (permutation P value < 0.001). The SBV method,
evaluated on the hold-out data has an AUC of 0.913 (95 % CI: 0.842–0.947). The
confidence intervals overlap such that we cannot find a difference in performance
between our model and the SBV model. Figure 4 shows the ROC curves of the
SBV and hexose uptake models. Our hexose uptake model is a decision tree with
a single feature:

```
reaction(A,1), link(A,B,_), link(B,C,_), reaction(C,1), reaction(B,1).
```

This corresponds to a chain of three *on* reactions, where the model predicts SSC if this feature exists and AC otherwise. This pathway activation pattern is present in 67 of the 76 individuals with SSC, and 17 of the 74 individuals with AC. In Fig. 4a we show an example instantiation of the hexose uptake pathway, for a particular individual. For this individual, the three variables A,B,C in the feature given above, are instantiated to the following reactions:

```
A. GLUT1 + ATP <=> GLUT1 :ATP.
B. GLUT1 + ATP <=> GLUT1 +ATP.
C. alpha-D-Glucose + ATP => alpha-D-glucose 6-Phosphate + ADP.
```

4.2   Results for Warmr Method

We illustrate the value of our Warmr only method using the cell cycle pathway (ranked fifth in Table 1). The more complex background predicates that we have defined for Warmr are only relevant when the pathway itself contains particular relationships. For example, the activation loop predicate will only be potentially beneficial when a pathway contains an activation edge, that may potentially be identified as the activation within an activation loop. The cell cycle is the top ranked pathway that contains all three kinds of edges; follows, activation and inhibition. The OneR classifier generated with the Warmr features has an AUC of 0.699 (95 % CI: 0.625–0.773), on the hold-out data.

While this model performs worse than the SBV Improver model and the hexose uptake pathway TreeLiker/combined model (in terms of AUC), it still has predictive value (Permutation P value < 0.001). The identified rule is complex and potentially interesting to a biologist:

```
actloop(C),largestlen(E,F,G),greaterthan(G,5),link(E,H,follows),r(H,0)
```

The rule states that a sample is classified as SCC cancer if there is a self activating loop for a reaction C, and that the longest chain of *on* reactions is from reaction E to reaction F, which is a chain at least 6 reactions long. Additionally, following reaction E there is also a reaction H that itself is not *on*.

This suggests that one of the differences between the SCC and AC cancer is that in the cell cycle SCC tumours have a self activating loop, that causes a longer chain of reactions to occur than in the AC tumour types. An instantiation

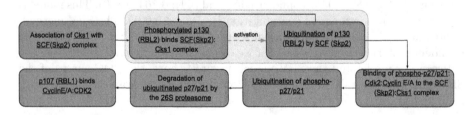

Fig. 5. The pattern found by Warmr instantiated for individual GSM1065725. There is a self-activating loop, highlighted by the grey box.

of this feature is shown in Fig. 5, for a particular individual. In this example there is a chain of 7 *on* reactions, and this also contains the self-activating loop.

4.3 Results for Warmr/TreeLiker Combined Method

As explained above, the feature used in the top pathway was very simple, and hence we demonstrate the value of our Warmr/TreeLiker combined approach on more complex features, identified from the hyaluronan biosynthesis and export pathway, ranked second in Table 1. Figure 6 shows the three features describing the subgroup identified by this approach. We can see that the additional edges that Warmr finds give a more complete view of the relations between the reactions in these features. This information may be important when a biologist analyses these results. The subgroup described by these three features has 58 true positives and 9 false positives in the hold-out data.

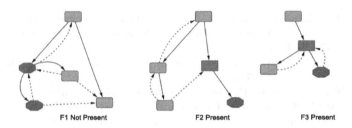

Fig. 6. The three features in the subgroup description. Solid lines represent the feature found by TreeLiker, dotted lines show the Warmr extensions. *on* reactions: green, rounded squares; *off*: red octagons; *on* or *off*: blue squares. (Color figure online)

5   Conclusions

In this work we have shown the potential of ILP methods for mining the abundance of highly structured biological data. Using this method we have identified differences in pathway activation patterns that go beyond the standard analysis of differentially expressed genes, enrichment analysis, gene feature ranking and pattern mining for common network motifs. We have also demonstrated the use of logical aggregation with a reaction graph, and how this simplifies the search for hypotheses to an extent that searching all pathways is tractable. We have introduced a novel approach that uses Warmr to extend features initially identified with TreeLiker. This makes it possible to search for long cyclical features.

We have identified pathway activation patterns predictive of the lung cancer type, in several pathways. The model we built on the hexose uptake pathway has predictive performance comparable with the top method from a recent challenge, but also provides biologically relevant explanations for its predictions. Each identified activation pattern is evaluated on the hold-out data, such that this should be the expected performance on new, unseen examples. The pathway activation patterns we have found are in clinically relevant pathways [28]. Patterns identified using this method may give diagnostic and clinical insights that biologists can develop into new hypotheses for further investigation.

Acknowledgments. LACM received funding from the Medical Research Council (MC_UU_12013/8).

References

1. Croft, D., Mundo, A.F., Haw, R., Milacic, M., Weiser, J., Guanming, W., Caudy, M., Garapati, P., Gillespie, M., Kamdar, M.R., et al.: The Reactome pathway knowledgebase. Nucleic Acids Res. **42**(D1), D472–D477 (2014)
2. Rhrissorrakrai, K., Jeremy Rice, J., Boue, S., Talikka, M., Bilal, E., Martin, F., Meyer, P., Norel, R., Xiang, Y., Stolovitzky, G., Hoeng, J., Peitsch, M.C.: SBV improver diagnostic signature challenge: design and results. Syst. Biomed. **1**(4), 3–14 (2013)
3. Tarca, A.L., Than, N.G., Romero, R.: Methodological approach from the best overall team in the SBV Improver Diagnostic Signature Challenge. Syst. Biomed. **1**(4), 217–227 (2013)
4. Draghici, S.: Statistical intelligence: effective analysis of high-density microarray data. Drug Discov. Today **7**(11), S55–S63 (2002)
5. Subramanian, A., Tamayo, P., Mootha, V.K., Mukherjee, S., Ebert, B.L., Gillette, M.A., Paulovich, A., Pomeroy, S.L., Golub, T.R., Lander, E.S., Mesirov, J.P.: Gene set enrichment analysis: A knowledge-based approach for interpreting genome-wide expression profiles. Proc. Nat. Acad. Sci. **102**(43), 15545–15550 (2005)
6. Gamberger, D., Lavrač, N., Železnỳ, F., Tolar, J.: Induction of comprehensible models for gene expression datasets by subgroup discovery methodology. J. Biomed. Inf. **37**(4), 269–284 (2004)
7. Holec, M., Klma, J., Železnỳ, F., Tolar, J.: Comparative evaluation of set-level techniques in predictive classification of gene expression samples. BMC Bioinform. **13**(Suppl 10), S15 (2012)
8. Whelan, K., Ray, O., King, R.D.: Representation, simulation, and hypothesis generation in graph and logical models of biological networks. In: Castrillo, J.I., Oliver, S.G. (eds.) Yeast Systems Biology, pp. 465–482. Springer, New York (2011)
9. Danon, L., Diaz-Guilera, A., Duch, J., Arenas, A.: Comparing community structure identification. J. Stat. Mech. Theory Exp. **2005**(09), P09008 (2005)
10. Kim, W., Li, M., Wang, J., Pan, Y.: Biological network motif detection and evaluation. BMC Syst. Biol. **5**(Suppl 3), S5 (2011)
11. Holec, M., Železnỳ, F., Kléma, J., Svoboda, J., Tolar, J.: Using bio-pathways in relational learning. Inductive Logic Programming, p. 50 (2008)
12. De Raedt, L.: Logical and Relational Learning. Springer Science & Business Media, New York (2008)
13. Flach, P.A., Lachiche, N.: 1BC: a first-order bayesian classifier. In: Džeroski, S., Flach, P.A. (eds.) ILP 1999. LNCS (LNAI), vol. 1634, pp. 92–103. Springer, Heidelberg (1999)
14. Lavrač, N., Vavpetič, A.: Relational and semantic data mining. In: Calimeri, F., Ianni, G., Truszczynski, M. (eds.) LPNMR 2015. LNCS, vol. 9345, pp. 20–31. Springer, Heidelberg (2015)

15. Dehaspe, L., De Raedt, L.: Mining association rules in multiple relations. In: Džeroski, S., Lavrač, N. (eds.) ILP 1997. LNCS, vol. 1297, pp. 125–132. Springer, Heidelberg (1997)

16. Ahmed, C.F., Lachiche, N., Charnay, C., Jelali, S.E., Braud, A.: Flexible propositionalization of continuous attributes in relational data mining. Expert Systems with Applications (2015)

17. Perlich, C., Provost, F.: Distribution-based aggregation for relational learning with identifier attributes. Mach. Learn. **62**(1–2), 65–105 (2006)

18. França, M.V.M., Zaverucha, G., d'Avila Garcez, A.S.: Fast relational learning using bottom clause propositionalization with artificial neural networks. Mach. Learn. **94**(1), 81–104 (2014)

19. Ristoski, P., Paulheim, H.: A comparison of propositionalization strategies for creating featuresfrom linked open data. In: Linked Data for Knowledge Discovery, p. 6 (2014)

20. Ristoski, P.: Towards linked open data enabled data mining. In: Gandon, F., Sabou, M., Sack, H., d'Amato, C., Cudré-Mauroux, P., Zimmermann, A. (eds.) ESWC 2015. LNCS, vol. 9088, pp. 772–782. Springer, Heidelberg (2015)

21. Kuželka, O., Železnỳ, F.: Block-wise construction of tree-like relational features with monotone reducibility and redundancy. Mach. Learn. **83**(2), 163–192 (2010)

22. Edgar, R., Domrachev, M., Lash, A.E.: Gene expression omnibus: NCBI gene expression and hybridizationarray data repository. Nucleic Acids Res. **30**(1), 207–210 (2002)

23. Wang, R.-S., Saadatpour, A., Albert, R.: Boolean modeling in systems biology: an overview of methodology and applications. Phy. Biol. **9**(5), 055001 (2012)

24. McCall, M.N., Jaffee, H.A., Zelisko, S.J., Sinha, N., Hooiveld, G., Irizarry, R.A., Zilliox, M.J.: The Gene Expression Barcode 3.0: improved data processing and mining tools. Nucleic Acids Res. **42**(D1), D938–D943 (2014)

25. Tyson, J.J., Chen, K.C., Novak, B.: Sniffers, buzzers, toggles and blinkers: dynamics of regulatory and signaling pathways in the cell. Curr. Opin. Cell Biol. **15**(2), 221–231 (2003)

26. Lavrač, N., Kavšek, B., Flach, P.A., Todorovski, L.: Subgroup discovery with CN2-SD. J. Mach. Learn. Res. **5**, 153–188 (2004)

27. Robin, X., Turck, N., Hainard, A., Tiberti, N., Lisacek, F., Sanchez, J.-C., Müller, M.: pROC: an open-source package for R and S+ to analyze and compare ROC curves. BMC Bioinform. **12**(1), 77 (2011)

28. Rongrong, W., Galan-Acosta, L., Norberg, E.: Glucose metabolism provide distinct prosurvival benefits to non-small cell lung carcinomas. Biochem. Biophy. Res. Commun. **460**(3), 572–577 (2015)

kProbLog: An Algebraic Prolog for Kernel Programming

Francesco Orsini[1,2]([✉]), Paolo Frasconi[2], and Luc De Raedt[1]

[1] Department of Computer Science, Katholieke Universiteit Leuven, Leuven, Belgium
{francesco.orsini,luc.deraedt}@cs.kuleuven.be
[2] Department of Information Engineering,
Universitá degli Studi di Firenze, Firenze, Italy
paolo.frasconi@unifi.it

Abstract. kProbLog is a simple algebraic extension of Prolog with facts and rules annotated with semiring labels. We propose kProbLog as a language for learning with kernels. kProbLog allows to elegantly specify systems of algebraic expressions on databases. We propose some code examples of gradually increasing complexity, we give a declarative specification of some matrix operations and an algorithm to solve linear systems. Finally we show the encodings of state-of-the-art graph kernels such as Weisfeiler-Lehman graph kernels, propagation kernels and an instance of Graph Invariant Kernels (GIKs), a recent framework for graph kernels with continuous attributes. The number of feature extraction schemas, that we can compactly specify in kProbLog, shows its potential for machine learning applications.

Keywords: Graph kernels · Prolog · Machine learning

1 Introduction

Statistical relational learning and probabilistic programming have contributed many declarative languages for supporting learning in relational representations. Prominent examples include Markov Logic [15], PRISM [16], Dyna [2] and ProbLog [1]. While these languages typically extend logical languages with probabilistic reasoning, there also exist extensions of a different nature: Dyna and aProbLog [9] are algebraic variations of probabilistic logical languages, while kLog [6] is a logical language for kernel-based learning.

Probabilistic languages such as PRISM and ProbLog label facts with probabilities, whereas Dyna and aProbLog use algebraic labels belonging to a semiring. Dyna has been used to encode many AI problems, including a simple distribution semantics, but does not support the disjoint-sum problem as ProbLog and aProbLog. While there has been a lot of research on integrating probabilistic and logic reasoning, kernel-based methods with logic have been much less investigated except for kLog and kFOIL [10]. kLog is a relational language for specifying kernel-based learning problems. It produces a graph representation of the learning problem in the spirit of knowledge-based model construction and

© Springer International Publishing Switzerland 2016
K. Inoue et al. (Eds.): ILP 2015, LNAI 9575, pp. 152–165, 2016.
DOI: 10.1007/978-3-319-40566-7_11

then employs a graph kernel on the resulting representation. kLog was designed to allow different graph kernels to be plugged in, but support to declaratively specify the exact form of the kernel is missing. kFOIL is a variation on the rule learner FOIL [14], that can learn kernels defined as the number of clauses that fire in both interpretations.

In the present paper, we investigate whether it is possible to use algebraic Prolog such as Dyna and aProbLog for kernel based learning. The underlying idea is that the labels will capture the kernel part, and the logic the structural part of the problem. Furthermore, unlike kLog and kFOIL, such a kernel based Prolog would allow to declaratively specify the kernel. More specifically, we propose kProbLog, a simple algebraic extension of Prolog, where kProbLog facts are labeled with semiring elements. kProbLog introduces meta-functions that allow to use different semirings in the same program and overcomes the limited expressiveness of semiring sum and product operations.

kProbLog can in principle handle the disjoint-sum problem as ProbLog and aProbLog, however in kernel design logical disjunctions and conjunctions are less common than algebraic sums and products, that are needed to specify matrix and tensor operations. We draw a parallel between kProbLog and tensors showing how to encode matrix operations in a way that is reminiscent of tensor relational algebra [8]. Nevertheless kProbLog supports recursion and so it is more expressive than tensor relational algebra.

We also show that kProbLog can be used for specifying or programming kernels on structured data in a declarative way. We use polynomials as kProbLog algebraic labels and show how they can be employed to specify label propagation and feature extraction schemas such as those used in recent graph kernels such as Weisfeiler-Lehman graph kernels [17], propagation kernels [12] and graph kernels with continuous attributes such as GIKs [13]. Polynomials were previously used in combination with logic programming for sensitivity analysis by Kimmig *et al.* [9] and for data provenance by Green *et al.* [7].

2   KProbLog^S

We propose kProbLog^S, an algebraic extension of Prolog in which facts and rules can be labeled with semiring elements.

Definition 1. *A kProbLog^S program P is a 4-tuple (F, R, S, ℓ) where:*

- *F is a finite set of facts,*
- *R is a finite set of definite clauses (also called rules),*
- *S is a semiring with sum $\oplus$ and product $\otimes$ operations, whose neutral elements are 0_S and 1_S respectively,*[1]
- *$\ell : F \to S$ is a function that maps facts to semiring values.*

---

[1] A semiring is an algebraic structure $(\mathcal{S}, \oplus, \otimes, 0_S, 1_S)$ where $\mathcal{S}$ is a set equipped with sum $\oplus$ and product $\otimes$ operations. Sum $\oplus$ and product $\otimes$ are associative and have as neutral element 0_S and 1_S respectively. The sum $\oplus$ is commutative, multiplication distributes w.r.t addition and 0_S is the annihilating element of multiplication.

We use the syntactic convention $\alpha::\mathtt{f}$ for algebraic facts where $f \in F$ is a fact and $\alpha = \ell(f)$ is the algebraic label.

Definition 2. *An algebraic interpretation $I_w = (I, w)$ of a ground kProbLog^S program P with facts F and atoms A is a set of tuples $(a, w(a))$ where a is an atom in the Herbrand base A and $w(a)$ is an algebraic formula over the fact labels $\{\ell(f)|f \in F\}$. We use the symbol $\emptyset$ to denote the empty algebraic interpretation, i.e. $\{(true, 1_S)\} \cup \{(a, 0_S)|a \in A\}$.*

We use an adaptation of the notation used by Vlasselaer *et al.* [18] in this definition and below.

Definition 3. *Let P be a ground algebraic logic program with algebraic facts F and atoms A. Let $I_w = (I, w)$ be an algebraic interpretation with pairs $(a, w(a))$. Then the $T_{(P,S)}$-operator is $T_{(P,S)}(I_w) = \{(a, w'(a))|a \in A\}$ where:*

$$
w'(a) = \begin{cases} \ell(a) & \text{if } a \in F \\ \displaystyle\bigoplus_{\substack{\{b_1,\ldots,b_n\}\subseteq I \\ a:-b_1,\ldots,b_n}} \bigotimes_{i=1}^{n} w(b_i) & \text{if } a \in A \setminus F \end{cases} . \tag{1}
$$

The least fixed point can be computed using a semi-naive evaluation. When the semiring is non-commutative the product $\otimes$ of the weights $w(b_i)$ must be computed in the same order that they appear in the rule. kProbLog^S can represent matrices that in principle can have infinite size and can be indexed by using elements of the Herbrand universe of the program. We now show some elementary kProbLog^S programs that specify matrix operations:

	algebra	kProbLog^S	numerical example

matrix A	A	`1::a(0, 0).`
		`2::a(0, 1).`
		`3::a(1, 1).`
			$\begin{bmatrix} 1 & 2 \\ 0 & 3 \end{bmatrix}$

matrix B	B	`2::b(0, 0).`
		`1::b(0, 1).`
		`5::b(1, 0).`
		`1::b(1, 1).`
			$\begin{bmatrix} 2 & 1 \\ 5 & 1 \end{bmatrix}$

matrix transpose	A^t	`c(I, J) :- a(J, I).`	$\begin{bmatrix} 1 & 2 \\ 0 & 3 \end{bmatrix}^t = \begin{bmatrix} 1 & 0 \\ 2 & 3 \end{bmatrix}$
matrix sum	$A + B$	`c(I, J) :- a(I, J).` `c(I, J) :- b(I, J).`	$\begin{bmatrix} 1 & 2 \\ 0 & 3 \end{bmatrix} + \begin{bmatrix} 2 & 1 \\ 5 & 1 \end{bmatrix} = \begin{bmatrix} 3 & 3 \\ 5 & 4 \end{bmatrix}$
matrix product	AB	`c(I, J) :-` `  a(I, K), b(K, J).`	$\begin{bmatrix} 1 & 2 \\ 0 & 3 \end{bmatrix}\begin{bmatrix} 2 & 1 \\ 5 & 1 \end{bmatrix} = \begin{bmatrix} 12 & 3 \\ 15 & 3 \end{bmatrix}$
Hadamard product	$A \odot B$	`c(I, J) :-` `  a(I, J), b(I, J).`	$\begin{bmatrix} 1 & 2 \\ 0 & 3 \end{bmatrix} \odot \begin{bmatrix} 2 & 1 \\ 5 & 1 \end{bmatrix} = \begin{bmatrix} 2 & 2 \\ 0 & 3 \end{bmatrix}$
Kronecker product	$\mathrm{kron}(A, B)$	`c(i(Ia, Ib), j(Ja, Jb)):-` `  a(Ia, Ja), b(Ib, Jb).`	$\begin{bmatrix} 1 & 2 \\ 0 & 3 \end{bmatrix} \otimes \begin{bmatrix} 2 & 1 \\ 5 & 1 \end{bmatrix} = \begin{bmatrix} 2 & 1 & 4 & 2 \\ 5 & 1 & 10 & 2 \\ 0 & 0 & 6 & 3 \\ 0 & 0 & 15 & 3 \end{bmatrix}$

The function symbols $\mathtt{i}/2$ and $\mathtt{j}/2$ were used to create the new indices that are needed by the Kronecker product. These definitions of matrix operations are reminiscent of tensor relational algebra [8]. Each of the above programs can be evaluated by applying the $T_{(P,S)}(I_w)$ operator only once. For each program we have a different definition of the C matrix that is represented by the predicate c/2. As a consequence of Eq. 1 all the algebraic labels of the c/2 facts are polynomials in the algebraic labels of the a/2 and b/2 facts. We draw an analogy between the representation of a sparse tensor in coordinate format and the representation of an algebraic interpretation. A ground fact can be regarded as a tuple of indices/domain elements that uniquely identifies the cell of a tensor, the algebraic label of the fact represents the value stored in the cell. In kProbLog^S for every atom a in the Herbrand base A the negation of a in an interpretation I_w can either be expressed with a sparse representation, by excluding it from the interpretation (i.e. $a \notin I_w$) or with a dense representation, including it in the interpretation with algebraic label 0_S (i.e. $a \in I_w$ and $w(a) = 0_S$).

Definition 4. *An algebraic interpretation $I_w = (I, w)$ is the fixed point of the $T_{(P,S)}(I_w)$-operator if and only if for all $a \in A$, $w(a) \equiv w'(a)$, where $w(a)$ and $w'(a)$ are algebraic formulae for a in I_w and $T_{(P,S)}(I_w)$ respectively.*

We denote with $T^i_{(P,S)}$ the function composition of $T_{(P,S)}$ with itself i times.

Corollary 1 (application of Kleene's theorem). *If S is an ω-continuous semiring[2] the algebraic system of fixed-point equations $I_w = T_{(P,S)}(I_w)$ admits a unique least solution $T^\infty_{(P,S)}(\emptyset)$ with respect to the partial order $\sqsubseteq$ and $T^\infty_{(P,S)}(\emptyset)$ is the supremum of the sequence $T^1_{(P,S)}(\emptyset), T^2_{(P,S)}(\emptyset), \ldots, T^i_{(P,S)}(\emptyset)$. So $T^\infty_{(P,S)}(\emptyset)$ can be approximated by computing successive elements of the sequence. If the semiring satisfies the ascending chain property (see [5]) then $T^\infty_{(P,S)}(\emptyset) = T^i_{(P,S)}(\emptyset)$ for some $i \geq 0$ and $T^\infty_{(P,S)}(\emptyset)$ can be computed exactly [5].*

We used $\emptyset$ to denote an empty algebraic interpretation. Examples of ω-continuous semirings are the boolean semiring ($\{T, F\}$, $\vee$, $\wedge$, F, T), the tropical semiring ($\mathbb{N} \cup \{\infty\}$, min, $+$, ∞, 0) and the fuzzy semiring ($[0, 1]$, max, min, 0, 1) [7]. Let us consider the following kProbLog^S program:

```
1::edge(a, b).          path(X, Y):-
3::edge(b, c).             edge(X, Y).
7::edge(a, c).          path(X, Y):-
                           edge(X, Z), path(Z, Y).
```

If S is the tropical semiring, we obtain a specification of the Floyd-Warshall algorithm for all-pair shortest paths on graphs. Assuming that S is the boolean semiring and all the algebraic labels that are different from 0_S correspond to true$\in S$, we obtain the Warshall algorithm for the transitive closure of a binary relation. Lehmann [11] explains how the Floyd-Warshall algorithm can be employed to

---

[2] A ω-continuous semiring is a naturally ordered semiring extended with an infinite summation-operator $\sum$. The natural order relation $\sqsubseteq$ on a semiring S is defined by $a \sqsubseteq b \Leftrightarrow \exists d \in S : a + d = b$. The semiring S is naturally ordered if $\sqsubseteq$ is a partial order on S; see [3,4] for details.

invert square matrices. The inverse A^{-1} of a square matrix A can be computed as the result of the transitive closure of $I - A$ where I is the identity matrix. The last example requires the capability to compute additive inverses which are not guaranteed to exist for semirings.

3   kProbLog

kProbLog generalizes kProbLog^S allowing multiple semirings and meta-functions. The coexistence of multiple semirings in the same program requires the declaration of the semiring of each algebraic predicate with the directive:

```
:- declare(<predicate>/<arity>, <semiring>).
```

We introduce meta-functions and meta-clauses to overcome the limits imposed by the semiring sum and product operations.

Definition 5 (meta-function). *A meta-function* m: $S_1 \times \ldots \times S_m \to S'$ *is a function that maps m semiring values $x_i \in S_i$, $i = 1, \ldots, k$ to a value of type S', where $S_1, \ldots, S_k$ and S' can be distinct semirings. Let* a_1,...,a_k *be algebraic atoms, the syntax* @m[a_1,...,a_k] *expresses that the meta-function* @m *is applied to the semiring values* w(a_1),...,w(a_k) *of the atoms* a_1,...,a_k.

Definition 6 (meta-clause). *In the kProbLog language a meta-clause* h :- b_1,...,b_n *is a universally quantified expression where* h *is an atom and* b_1,...,b_n *can be either body atoms or meta-functions applied to other algebraic atoms. For a given meta-clause, if the head is labeled with the semiring S, also the labels of the body atoms and the return types of the meta-functions must be on the semiring S.*

Definition 7 (kProbLog program). *A kProbLog program P is a union of kProbLog^{S_i} programs and meta-clauses.*

The introduction of meta-functions in kProbLog allows us to deal with other algebraic structures such as rings that require the additive inverse @minus/1 and fields that require the additive inverse and the multiplicative inverse @inv/1.

3.1   Recursive kProbLog Program with Meta-Functions

kProbLog allows both additive and destructive updates, the update type is specified by the directive:

```
:- declare(<predicate>/<arity>, <semiring>, <update-type>).
```

where update-type can be either additive or destructive.[3]

---

[3] The directive declare/3 must be used instead of declare/2 whenever the groundings of the declared predicate appear in the cycles of the ground program. In case the directive declare/3 is not specified this can be detected by the system at evaluation time.

We propose a simple example of an algebraic program that uses meta-functions to compute the limit $\lim_{n \to +\infty} g^n(x_0)$ of an iterated function $g(x) = x(1 - x)$, where $g^n(x_0)$ of the function-composition of g with itself n times starting from some initial number x_0.

Assuming that $x_0 = 0.5$, in kProbLog we would write:

```
:- declare(x, real, destructive).
:- declare(x0, real).
0.5::x0.
x :- x0.
x :- @g[x].
```

The above program has the following behaviour: the weight $w(x)$ of x is initialised to $w(x_0) = 0.5$ and then updated at each step according to the rule $w'(x) = g(w(x))$. The directive :- declare(x, real, destructive). causes the result of the immediate-consequence operator to be used as a destructive assignment of the weights instead of an additive update.

We could have also considered an additive update rule such as $w'(x) = w(x) + g(w(x))$, but this would not lead us to the expected result for an iterated function system.

While iterated function systems require an update with destructive assignment, other programs such as the transitive closure of a binary relation (see above) or the compilation of ProbLog programs with SDDs require additive updates.[4]

3.2   The Jacobi Method

We already showed that kProbLog can express linear algebra operations, we now combine recursion and meta-functions in an algebraic program that specifies the Jacobi method. The Jacobi method is an iterative algorithm used for solving diagonally dominant systems of linear equations $Ax = b$.

We consider the field of real numbers $\mathbb{R}$ (i.e. kProbLog$^{\mathbb{R}}$) as semiring together with the meta-functions @minus and @inv that provide the inverse element of sum and product respectively.

The A matrix must be split according to the Jacobi method:

$$D = \text{diag}(A) \qquad \texttt{d(I, I) :- a(I, I).}$$
$$R = A - D \qquad \texttt{r(I, J) :- a(I, J), I \textbackslash= J.}$$

The solution $\mathbf{x}^*$ of $A\mathbf{x} = \mathbf{b}$ is computed iteratively by finding the fixed point of $\mathbf{x} = D^{-1}(\mathbf{b} - R\mathbf{x})$. We call E the inverse of D. Since D is diagonal also E is a diagonal matrix:

$$e_{ii} = \text{invert}(d_{ii}) = \frac{1}{d_{ii}} \qquad \texttt{e(I, I) :- @invert[d(I, I)].}$$

and the iterative step can be rewritten as $\mathbf{x} = E(\mathbf{b} - R\mathbf{x})$.

---

[4] The compilation of ProbLog programs [18] can be expressed in kProbLog, provided that the SDD semiring is used. The update of the algebraic weights must be additive, each update adds new proves for the ground atoms until convergence.

Making the summations explicit we can write:

$$x_i = \sum_k e_{ik} \left(b_k - \sum_l r_{kl} x_l \right) \tag{2}$$

then we can extrapolate the term $\sum_l r_{kl} x_l$ turning it into the aux_k definition:

$$x_i = \sum_k e_{ik} (b_k - aux_k)$$
$$aux_k = \sum_l r_{kl} x_l$$

```
:- declare(x/1, real, destructive).
:- declare(aux/1, real, destructive).
x(I) :-
    e(I, K), @subtraction[b(K), aux(K)].

aux(I) :-
    r(K, L),  x(L).
```

where @subtraction/2 represents the subtraction between real numbers, x/1 and aux/1 are mutually recursive predicates. Because x/1 needs to be initialized (perhaps at random) we also need the clause:

$$x_i = init_i \qquad\qquad \texttt{x(I) :- init(I).}$$

where init/1 is a unary predicate. This example also shows that kProbLog is more expressive than tensor relational algebra because it supports recursion.

3.3   kProbLog T_P-Operator with Meta-Functions

The algebraic T_P-operator of kProbLog is defined on the meta-transformed program.

Definition 8 (meta-transformed program). *A meta-transformed kProbLog program is a kProbLog program in which all the meta-functions are expanded to algebraic atoms. For each rule* h :- b_1,...,@m[a_1,...,a_k],...,b_n *in the program P each meta-function* @m[a_1,...,a_k] *is replaced by a body atom* b' *and a meta-clause* b':-@m[a_1,...,a_k] *is added to the program P.*

Definition 9 (algebraic T_P-operator with meta-functions). *Let P be meta-transformed kProbLog program with facts F and atoms A. Let $I_w = (I, w)$ be an algebraic interpretation with pairs $(a, w(a))$. Then the T_P-operator is $T_P(I_w) = \{(a, w'(a)) | a \in A\}$ where:*

$$w'(a) = \begin{cases} \ell(a) & \text{if } a \in F \\ \displaystyle\bigoplus_{\substack{\{b_1,...,b_n\} \subseteq I \\ a:-b_1,...,b_n}} \bigotimes_{i=1}^n w(b_i) \oplus \bigoplus_{\substack{\{b_1,...,b_k\} \subseteq I \\ a:-@m[b_1,...,b_k]}} m(w(b_1),...,w(b_k)) & \text{if } a \in A \setminus F. \end{cases} \tag{3}$$

The introduction of meta-functions makes the result of the evaluation of a kProbLog program dependent on the order in which rules and meta-clauses are evaluated. For this reason we explain the order adopted by the kProbLog language. A kProbLog program P is grounded to a program ground(P) and then partitioned into a sequence of strata $P_1, \ldots, P_n$.

An atom in a non-recursive stratum P_i can only depend on the atoms from the previous strata $\bigcup_{j<i} P_j$, while an atom in a recursive stratum can depend on the atoms in $\bigcup_{j\leq i} P_j$.[5] Each partition P_i must be maximal and strongly connected (i.e. each atom in P_i depends on every other atom in P_i). The program evaluation starts by initializing the weight $w(a)$ of each ground atom a in ground(P) with 0_S where S is the semiring of the atom. Then the strata are visited in order and the weights are updated as follows: if the stratum P_i is non-recursive we apply the algebraic T_P-operator only once per atom, while if P_i is recursive we apply the algebraic T_P-operator only once for the non-recursive rules and meta-clauses and repeatedly until convergence for the recursive rules and meta-clauses.

When updating the weight $w(a)$ of a recursive atom a at each iteration we initialize a weight $\Delta w(a) = 0_s$. We accumulate on $\Delta w(a)$ the result of the application of the T_P-operator on all the recursive rules with head a. Then the new weight for a is computed as $w(a) = w(a) + \Delta w(a)$ or $w(a) = \Delta w(a)$ for additive and destructive update respectively.

If P_i is a cyclic stratum then the convergence of the algebraic T_P-operator must be guaranteed by the user that specifies the program. Nevertheless if the P_i is a cyclic stratum in which only rules are cyclic all the atoms in P_i are on the same semiring[6] S and so P_i has the same convergence properties of a kProbLog^S program (see Corollary 1 on page 4). Whenever we apply the algebraic T_P-operator we use the Jacobi evaluation, so that the program is not affected by the order in which rules and meta-clauses are evaluated. This program evaluation procedure is an adaptation the work of Whaley et al. [19] on Datalog and binary decision diagrams.

4   kProbLog$^{S[\mathsf{x}]}$

kProbLog$^{S[\mathsf{x}]}$ labels facts and rule heads with polynomials over the semiring S. kProbLog$^{S[\mathsf{x}]}$ is a particular case of kProbLog^S because polynomials over semirings are semirings in which addition and multiplication are defined as usual.

Definition 10 (Multivariate polynomials over commutative semirings). *A multivariate polynomial* $\mathcal{P} \in S[\mathbf{x}]$ *can be expressed as:*

$$\mathcal{P}(\mathbf{x}) = \bigoplus_{i=1}^{n} c_i \mathbf{x}^{\mathbf{e}_i} = \bigoplus_{i=1}^{n} c_i \otimes \bigotimes_{t \in T_i} x_t^{e_{it}} \qquad (4)$$

---

[5] We say that an atom **a** *directly depends* on an atom **b** if **a** is the head of a rule or a meta-clause and **b** is a body literal or an argument of a meta-function in the meta clause. We say that an atom **a** *depends* on an atom **b** either if **a** directly depends on **b** or there is an atom **c** such that **a** directly depends on **c** and **c** depends on **b**.

[6] Atoms of distinct semirings cannot be mutually dependent without using meta-clauses.

where $c_i \in S$ are the coefficients of the ith monomial and $\mathbf{x}$, $\mathbf{e}$ are vectors of variables and exponents respectively. The vector $\mathbf{x}$ is indexed by ground terms $t \in T$.

4.1 Polynomials for Feature Extraction

We shall use polynomials to represent kernel features such as the ones computed by the Weisfeiler-Lehman and propagation kernels. We define an inner-product between multivariate polynomials of $\mathbb{R}[\mathbf{x}]$, with a finite number of monomials as:

$$\langle \mathcal{P}(\mathbf{x}), \mathcal{Q}(\mathbf{x}) \rangle = \sum_{(p,e)\in\mathcal{P}} \sum_{(q,e)\in\mathcal{Q}} pq. \tag{5}$$

For each monomial (uniquely identified by the vector of exponents $\mathbf{e}$) that appears in both the polynomials $\mathcal{P}$ and $\mathcal{Q}$, Eq. 5 computes the product between their coefficients p and q respectively. These products are then summed together to obtain the value of the inner-product.

For example we can consider the multivariate polynomials on integer coefficients:

$$\begin{aligned} \mathcal{P}(x_1, x_2, x_3) &= 2x_1 + 3x_1x_2 + x_2x_3^2 \\ \mathcal{Q}(x_1, x_2, x_3) &= 4x_1 + 3x_1x_3 + 3x_2x_3^2 \end{aligned} \tag{6}$$

which can be expressed as two sets of coefficient-exponent pairs $\mathcal{P} = \{(2, [1, 0, 0]), (3, [1, 1, 0]), (1, [0, 1, 2])\}$ and $\mathcal{Q} = \{(4, [1, 0, 0]), (3, [1, 0, 1]), (3, [0, 1, 2])\}$ respectively. The two polynomials have in common the vectors of exponents $[1, 0, 0]$ and $[0, 1, 2]$, each contributes to the inner product by $2 \times 4 = 8$ and $1 \times 3 = 3$ respectively. The value of the inner product between $\mathcal{P}(x_1, x_2, x_3)$ and $\mathcal{Q}(x_1, x_2, x_3)$ is the sum of such contributes $8 + 3 = 11$.

In kProbLog the inner-product between two algebraic atoms $\mathcal{P}(\mathbf{x})\text{::a}$ and $\mathcal{Q}(\mathbf{x})\text{::b}$ can be computed using the meta-function @dot/2. Another meta-function, that is useful for kernel design, is @rbf/3. The meta-function @rbf/3 takes as input an atom labeled with a non-negative real value γ and two atoms labeled with the polynomials $\mathcal{P}$ and $\mathcal{Q}$ and computes the rbf kernel $exp\{-\gamma\|\mathcal{P} - \mathcal{Q}\|^2\}$.[7]

4.2 The @id Meta-Function

The @id/1 meta-function @id: $S \to S$ is injective and transforms a polynomial $\mathcal{P}(\mathbf{x})$ to a new term t and returns the polynomial $\text{@id}[\mathcal{P}(\mathbf{x})] = 1.0 \cdot x(t)$. This function can be used to compress a multivariate polynomial to a new polynomial in a single variable. We use the @id meta-function for polynomial compression as Shervashidze *et al.* [17] use the function f to compress multisets of labels.

Indeed we can represent a multiset μ of labels (we use Prolog ground terms to represent labels) as a polynomial:

$$\mathcal{P}_\mu(\mathbf{x}) = \sum_{t\in\mu} \sharp t \cdot x(t) \tag{7}$$

---

[7] The squared distance in the rbf kernel can be expressed by using the dot product, i.e. $\|\mathcal{P} - \mathcal{Q}\|^2 = \langle \mathcal{P}, \mathcal{P} \rangle + \langle \mathcal{Q}, \mathcal{Q} \rangle - 2\langle \mathcal{P}, \mathcal{Q} \rangle$.

where $\sharp$ counts the number of occurrences of the label (identified by the ground term t) in the multiset μ.

Weisfeiler-Lehman algorithm: A colored graph G is a triple (V, E, ℓ) where V is a set of vertices, $E \subseteq V \times V$ is the set of the edges and $\ell : V \to \Sigma$ is a function that maps vertices to a color alphabet Σ. For example we can specify vertex labels and edge connectivity of a graph graph_a in kProbLog as follows:

```
:- declare(vertex/2, polynomial(int)).    1.0::edge(Graph, A, B):-
:- declare(edge_asymm/3, boolean).            edge_asymm(Graph, A, B).
:- declare(edge/3, polynomial(int)).
                                           1.0::edge(Graph, A, B):-
1 * x(pink)::vertex(graph_a, 1).              edge_asymm(Graph, B, A).
1 * x(blue)::vertex(graph_a, 2).
1 * x(blue)::vertex(graph_a, 3).
1 * x(blue)::vertex(graph_a, 4).
1 * x(blue)::vertex(graph_a, 5).

edge_asymm(graph_a, 1, 2).
edge_asymm(graph_a, 1, 3).
edge_asymm(graph_a, 2, 4).
edge_asymm(graph_a, 3, 4).
edge_asymm(graph_a, 4, 5).
```

where the boolean predicate edge_asymm/3 is implicitly casted to integer and then to polynomial over integers when it appears in the definition of edge/3. The Weisfeiler-Lehman color of a vertex after h steps of the algorithm is defined as:

$$\mathcal{L}^h(v) = \begin{cases} \ell(v) & \text{if } h = 0 \\ f(\{\mathcal{L}^{h-1}(w)|w \in \mathcal{N}(v)\}) & \text{if } h > 0 \end{cases} \tag{8}$$

where $\mathcal{N}(v)$ is the set of the vertex neighbors of v an $\{\mathcal{L}^{h-1}(w)|w \in \mathcal{N}(v)\}$ is the multiset of their colors at step $h - 1$. The Weisfeiler-Lehman algorithm can be specified in kProbLog using the recursive definition of Eq. 8:

```
:- declare(wl_color/3,              wl_color(0, Graph, V):-
     polynomial(int)).                vertex(Graph, V).
:- declare(wl_color_multiset/3,
     polynomial(int)).              wl_color(H, Graph, V):-
                                      H > 0,
wl_color_multiset(H, Graph, V):-      H1 is H - 1,
  edge(Graph, V, W),                  @id[wl_color_multiset(H1, Graph, V)].
  wl_color(H, Graph, W).
```

5 Graph Kernels

In this section we give the declarative specification of some recent graph kernels such as the Weisfeiler-Lehman graph kernel [17], propagation kernels [12] and graph invariant kernels [13]. These methods have been applied to different domains such as: natural language processing [13], computer vision [12] and bioinformatics [12,13,17].

5.1   Weisfeiler-Lehman Graph Kernel and Propagation Kernels

The Weisfeiler-Lehman graph kernel is defined using a base kernel [17] that computes the inner-product between the histograms of Weisfeiler-Lehman colors of two graphs Graph and GraphPrime.

```
:- declare(phi/2, real).          :- declare(base_kernel/3, real).
phi(H, Graph):-                    base_kernel(H, Graph, GraphPrime):-
   wl_color(H, Graph, V).            @dot[phi(H, Graph),
                                          phi(H, GraphPrime)].
```

The Weisfeiler-Lehman graph kernel [17] with H iterations is the sum of base kernels computed for consecutive Weisfeiler-Lehman labeling steps 1,...,H on the graphs Graph and GraphPrime:

```
:- declare(kernel_wl/3, real).
kernel_wl(0, Graph, GraphPrime):-    kernel_wl(H, Graph, GraphPrime):-
   base_kernel(0, Graph, GraphPrime).    H > 0,
                                          base_kernel(H, Graph, GraphPrime).
kernel_wl(H, Graph, GraphPrime):-       .
   H > 0, H1 is H - 1,
   kernel_wl(H1, Graph, GraphPrime).
```

Propagation kernels [12] are a generalization of the Weisfeiler-Lehman graph kernel, that can adopt different label propagation schemas. Neumann *et al.* [12] implements propagation kernels using locality sensitive hashing. The kProbLog specification is identical to the one the Weisfeiler-Lehman except that the @id meta-function is to be replaced with a meta-function that does locality sensitive hashing.

5.2   Graph Invariant Kernels

Graph Invariant Kernels (GIKs, pronounce "geeks") are a recent framework for graph kernels with continuous attributes [13]. GIKs compute a similarity measure between graphs G and G' matching them at vertex level according to the formula:

$$k(G, G') = \sum_{v \in V(G)} \sum_{v' \in V(G')} w(v, v') k_{\text{ATTR}}(v, v') \tag{9}$$

where $w(v, v')$ is the structural weight matrix and $k_{\text{ATTR}}(v, v')$ is a kernel on the continuous attributes of the graphs. We use R-neighborhood subgraphs, so the kProbLog specification is parametrized by the variable R.

```
:- declare(gik_radius/3, real).
gik_radius(R, Graph, GraphPrime):-
   w_matrix(R, Graph, V, GraphPrime, VPrime),
   k_attr(Graph, V, GraphPrime, VPrime).
```

where gik_radius/3, w_matrix/5 and k_attr/4 are algebraic predicates on the real numbers semiring, which is represented with floats for implementation purposes. Assuming that we want to use the RBF with $\gamma = 0.5$ kernel on the vertex attributes we can write:

```
:- declare(rbf_gamma_const/0, real).
:- declare(k_attr/4, real).
0.5::rbf_gamma_const.
k_attr(Graph, V, GraphPrime, VPrime):-
    @rbf[rbf_gamma_const, attr(Graph, V), attr(GraphPrime, VPrime)].
```

where `attr/2` is an algebraic predicate that associates to the vertex V of a Graph a polynomial label. To associate to vertex v_1 of graph_a the 4-dimensional feature $[1, 0, 0.5, 1.3]$ we would write:

```
:- declare(attr/2, polynomial(real)).
1.0 * x(1) + 0.5 * x(3) + 1.3 * x(4)::attr(graph_a, v_1).
```

while the meta-function `@rbf/3` takes as input an atom `rbf_gamma_const` labeled with the γ constant and the atoms relative to the vertex attributes.

The structural weight matrix $w(v, v')$ is defined as:

$$w(v, v') = \sum_{g \in \mathcal{R}^{-1}(G)} \sum_{g' \in \mathcal{R}^{-1}(G')} k_{\text{INV}}(v, v') \frac{\delta_m(g, g')}{|V_g||V_{g'}|} \mathbb{1}\{v \in V_g \wedge v' \in V_{g'}\}. \quad (10)$$

The weight $w(v, v')$ measures the structural similarity between vertices and is defined combining an $\mathcal{R}$-decomposition relation, a function $\delta_m(g, g')$ and a kernel on vertex invariants k_{INV} [13]. In our case the $\mathcal{R}$-decomposition generates R-neighborhood subgraphs (the same used in the experiments of Orsini *et al.* [13]).

There are multiple ways to instantiate GIKs, we choose the version called LWL$_V$, because as shown with the experiments by Orsini *et al.* [13], can achieve very good accuracies most of the times. LWL$_V$ uses R-neighborhood subgraphs $\mathcal{R}$-decomposition relation, computes the kernel on vertex invariants $k_{\text{INV}}(v, v')$ at the pattern level (*local* GIK) and uses $\delta_m(g, g')$ to match subgraphs that have the same number of nodes.

In kProbLog we would write:

```
:- declare(w_matrix/5, real).
w_matrix(R, Graph, V, GraphPrime, VPrime):-
    vertex_in_ball(Graph, R, BallRoot, V),
    vertex_in_ball(GraphPrime, R, BallRootPrime, VPrime),
    delta_match(R, Graph, BallRoot, GraphPrime, BallRootPrime),
    @inv[ball_size(R, Graph, BallRoot)],
    @inv[ball_size(R, GraphPrime, BallRootPrime)],
    k_inv(Graph, BallRoot, V, GraphPrime, BallRootPrime, VPrime).
```

where:
(a) `vertex_in_ball(R, Graph, BallRoot, V)` is a boolean predicate which is true if V is a vertex of Graph inside a R-neighborhood subgraph rooted in BallRoot. `vertex_in_ball/4` encodes both the term $\mathbb{1}\{v \in V_g \wedge v' \in V_{g'}\}$ and the pattern generation of the decomposition relation $g \in \mathcal{R}^{-1}(G)$.

```
:- declare(vertex_in_ball/4, bool).        vertex_in_ball(R, Graph, Root, V):-
vertex_in_ball(0, Graph, Root, Root):-       R > 0, R1 is R - 1,
    vertex(Graph, Root).                      edge(Graph, Root, W),
                                              vertex_in_ball(R1, Graph, W, V).
vertex_in_ball(R, Graph, Root, V):-
    R > 0, R1 is R - 1,
    vertex_in_ball(R1, Graph, Root, V).
```

(b) delta_match(R, Graph, BallRoot, GraphPrime, BallRootPrime)
matches subgraphs with the same number of vertices

```
:- declare(delta_match/5, real).
:- declare(v_id/3, polynomial(real)).
:- declare(ball_size/3, int).
delta_match(R, Graph, BallRoot, GraphPrime, BallRootPrime):-
  @eq[v_id(R, Graph, BallRoot), v_id(R, GraphPrime, BallRootPrime)].

v_id(R, Graph, BallRoot):- @id[ball_size(R, Graph, BallRoot)].

ball_size(R, Graph, BallRoot):- vertex_in_ball(R, Graph, BallRoot, V).
```

(c) @inv[ball_size(Radius, Graph, BallRoot)] corresponds to the normalization term $1/|V_g|$. @inv is the meta-function that computes the multiplicative inverse and ball_size(Radius, Graph, BallRoot) is a the float predicate that counts the number of vertices in a Radius-neighborhood rooted in BallRoot.

(d) k_inv(R, Graph, BallRoot, V, GraphPrime, BallRootPrime, VPrime) computes k_{INV} using H_WL iterations of the Weisfeiler-Lehman algorithm to obtain vertex features phi_wl(R, H_WL, Graph, BallRoot, V) from the R-neighborhood subgraphs.

```
:- declare(k_inv/7, real).
:- declare(phi_wl/5, polynomial(real)).
wl_iterations(3). % constant

k_inv(R, Graph, BallRoot, V, GraphPrime, BallRootPrime, VPrime):-
  wl_iterations(H_WL),
  @dot[phi_wl(R, H_WL, Graph, BallRoot, V),
    phi_wl(R, H_WL, GraphPrime, BallRootPrime, VPrime)].

phi_wl(R, 0, Graph, BallRoot, V):-      phi_wl(R, H, Graph, BallRoot, V):-
  wl_color(R, Graph, BallRoot, 0, V).     H > 0, H1 is H-1,
                                          phi_wl(R, H1, Graph, BallRoot, V).
phi_wl(R, H, Graph, BallRoot, V):-
  H > 0, wl_color(R, Graph, BallRoot, H, V).
```

where wl_color/5 is defined as wl_color/3, but has two additional arguments R and BallRoot that are needed to restrict the graph connectivity to the R-neighborhood subgraph rooted in vertex BallRoot.

6   Conclusions

We proposed kProbLog, a simple algebraic extension of Prolog that can be used for kernel programming. Polynomials and meta-functions allow to elegantly specify in kProbLog many recent kernels (e.g. the Weisfeiler-Lehman Graph kernel, propagation kernels and GIKs). kProbLog rules are used for kernel programming, but also to incorporate background knowledge and enrich the input data representation with user specified relations. kProbLog is a language that provides a uniform representation for relational data, background knowledge and kernel design. In our future work we will exploit these three characteristics of kProbLog to learn feature spaces with inductive logic programming.

References

1. De Raedt, L., Kimmig, A., Toivonen, H.: Problog: A probabilistic prolog and its application in link discovery. In: IJCAI (2007)
2. Eisner, J., Filardo, N.W.: Dyna: extending datalog for modern AI. In: de Moor, O., Gottlob, G., Furche, T., Sellers, A. (eds.) Datalog 2010. LNCS, vol. 6702, pp. 181–220. Springer, Heidelberg (2011)
3. Esparza, J., Luttenberger, M.: Solving fixed-point equations by derivation tree analysis. In: Corradini, A., Klin, B., Cîrstea, C. (eds.) CALCO 2011. LNCS, vol. 6859, pp. 19–35. Springer, Heidelberg (2011)
4. Esparza, J., Kiefer, S., Luttenberger, M.: An extension of newton's method to ω-continuous semirings. In: Harju, T., Karhumäki, J., Lepistö, A. (eds.) DLT 2007. LNCS, vol. 4588, pp. 157–168. Springer, Heidelberg (2007)
5. Esparza, J., Luttenberger, M., Schlund, M.: FPSOLVE: a generic solver for fixpoint equations over semirings. In: Holzer, M., Kutrib, M. (eds.) CIAA 2014. LNCS, vol. 8587, pp. 1–15. Springer, Heidelberg (2014)
6. Frasconi, P., Costa, F., De Raedt, L., De Grave, K.: kLog: A language for logical and relational learning with kernels. In: Artificial Intelligence (2014)
7. Green, T.J., Karvounarakis, G., Tannen, V.: Provenance semirings. In: Proceedings of the 26th ACM SIGMOD-SIGACT-SIGART Symposium on Principles of database systems. ACM (2007)
8. Kim, M., Candan, K.S.: Approximate tensor decomposition within a tensor-relational algebraic framework. In: Proceedings of the 20th ACM International Conference on Information and Knowledge Management. ACM (2011)
9. Kimmig, A., Van den Broeck, G., De Raedt, L.: An algebraic prolog for reasoning about possible worlds. In: 25th AAAI Conference on Artificial Intelligence (2011)
10. Landwehr, N., Passerini, A., De Raedt, L., Frasconi, P.: kFOIL: Learning simple relational kernels. In: AAAI (2006)
11. Lehmann, D.J.: Algebraic structures for transitive closure. In: Theoretical Computer Science (1977)
12. Neumann, M., Patricia, N., Garnett, R., Kersting, K.: Efficient graph kernels by randomization. In: Flach, P.A., Bie, T., Cristianini, N. (eds.) ECML PKDD 2012, Part I. LNCS, vol. 7523, pp. 378–393. Springer, Heidelberg (2012)
13. Orsini, F., Frasconi, P., De Raedt, L.: Graph invariant kernels. In: Proceedings of the 24th IJCAI (2015)
14. Ross Quinlan, J.: Learning logical definitions from relations. Mach. Learn. 5(3), 239–266 (1990)
15. Richardson, M., Domingos, P.: Markov logic networks. Machine Learning (2006)
16. Sato, T., Kameya, Y.: PRISM: a language for symbolic-statistical modeling. In: IJCAI (1997)
17. Shervashidze, N., Schweitzer, P., Van Leeuwen, E.J., Mehlhorn, K., Borgwardt, K.M.: Weisfeiler-lehman graph kernels. J. Mach. Learn. Res. 12, 2539–2561 (2011)
18. Vlasselaer, J., Van den Broeck, G., Kimmig, A., Meert, W., De Raedt, L.: Anytime inference in probabilistic logic programs with tp-compilation. In: Proceedings of the 24th IJCAI (2015)
19. Whaley, J., Avots, D., Carbin, M., Lam, M.S.: Using datalog with binary decision diagrams for program analysis. In: Yi, K. (ed.) APLAS 2005. LNCS, vol. 3780, pp. 97–118. Springer, Heidelberg (2005)

An Exercise in Declarative Modeling
for Relational Query Mining

Sergey Paramonov[✉], Matthijs van Leeuwen, Marc Denecker,
and Luc De Raedt

KU Leuven, Celestijnenlaan 200A, 3001 Heverlee, Belgium
{sergey.paramonov,matthijs.leeuwen,marc.denecker,
luc.raedt}@cs.kuleuven.be

Abstract. Motivated by the declarative modeling paradigm for data
mining, we report on our experience in modeling and solving relational
query and graph mining problems with the IDP system, a variation
on the answer set programming paradigm. Using IDP or other ASP-
languages for modeling appears to be natural given that they provide
rich logical languages for modeling and solving many search problems
and that relational query mining (and ILP) is also based on logic. Nevertheless, our results indicate that second order extensions to these languages are necessary for expressing the model as well as for efficient
solving, especially for what concerns subsumption testing. We propose
such second order extensions and evaluate their potential effectiveness
with a number of experiments in subsumption as well as in query mining.

Keywords: Knowledge representation · Answer set programming ·
Data mining · Query mining · Pattern mining

1   Introduction

In the past few years, many pattern mining problems have been modeled using
constraint programming techniques [5]. While the resulting systems are not
always as efficient as state-of-the-art pattern mining systems, the advantages
of this type of declarative modeling are now generally accepted: they support
more constraints, they are easier to modify and extend, and they are built using
general purpose systems. However, so far, the declarative modeling approach
has not yet been applied to inductive logic programming. This paper investi-
gates whether such an extension would be possible. To realize this, we consider
frequent query mining, the ILP version of frequent pattern mining, as well as the
answer set programming paradigm, the logic programming version of constraint
programming. More specifically, we address the following three questions:

Q_1 Is it possible to design and implement a declarative query miner that uses a
logical and relational representation for both the data and the query mining
problem?

© Springer International Publishing Switzerland 2016
K. Inoue et al. (Eds.): ILP 2015, LNAI 9575, pp. 166–182, 2016.
DOI: 10.1007/978-3-319-40566-7_12

Q₂ Is it possible to take advantage of recent progress in the field of computational logic by adopting an Answer Set Programming (ASP) [23] framework for modeling and solving?

Q₃ Would such a system be computationally feasible? That is, can it tackle problems of at least moderate size?

Our study is not only relevant to ILP, but also to the field of knowledge representation and ASP as query mining (and ILP) is a potentially interesting application that may introduce new challenges and suggest solver extensions.

More concretely, the main contributions of this work can be summarized as follows:

1. We present two declarative models and corresponding solving strategies for the query mining problem that support a wide variety of constraints. While one model can be expressed in the ASP paradigm, the other model requires a second order extension that we believe to be essential for modeling ILP tasks.
2. We implement and evaluate the presented models in the IDP system [2], a knowledge base system that belongs to the ASP paradigm.
3. We empirically evaluate the proposed models and compare them on the classical datasets with the state-of-the-art ILP methods.

This paper is organized as follows: Sect. 2 formally introduces the problem. In Sect. 3, we introduce a second order model for frequent query mining that addresses Question Q_1. In Sect. 4 we present a first order model for query mining, demonstrate main issues with this approach and address Question Q_2. In Sect. 5 we provide experimental evidence to support our answers to Questions Q_2 and Q_3. In Sect. 6 we discuss advantages (such as extendability) and disadvantages of the models and the approach overall. In Sect. 7 we present an overview of the related work in the ILP context of frequent query mining. Finally, we conclude in Sect. 8 with a summary of the work.

2 Problem Statement

The problem that we address in this paper is to mine queries in a logical and relational learning setting. Starting with the work the Warmr system [10], there has been a line of work that focusses on the following frequent query mining problem [8, 16, 19]:

Given:

- a relational database D,
- the entity of interest determining the $key/1$ predicate,
- a frequency threshold t,
- a language $\mathcal{L}$ of logical queries of the form $key(X) \leftarrow b_1, ..., b_n$ defining $key/1$ (b_i's are atoms).

Find: all queries $c \in \mathcal{L}$ s.t. $freq(c, D) \geq t$, where $freq(c, D) = |\{\theta \mid D \cup c \models key(X)\theta\}|$.

Notice that the substitutions θ only substitute variables X that occur in *key*.

In this paper, we focus our attention on graph data, as this forms the simplest truly relational learning setting and allows us to focus on what is essential for extending the declarative modeling paradigm to a relational setting. In principle, this setting can easily be generalized to the full inductive logic programming problem.

As an example, consider a graph database D, represented by the facts

$$\{edge(g_1, 1, 2),\ edge(g_1, 2, 3),\ edge(g_1, 1, 3),\ edge(g_2, 1, 2),\ edge(g_2, 2, 3),\ edge(g_2, 1, 3), \dots\},$$

where the ternary relation $edge(g, e_1, e_2)$ states that in graph g there is an edge between e_1 and e_2 (we assume graphs to be undirected, so there is also always an edge between e_2 and e_1). The frequency of $key(K) \leftarrow edge(K, B, C)$, $edge(K, C, D), edge(K, B, D)$ in this database is 2 as the query returns g_1 and g_2. If $key(g)$ holds, then the graph g is subsumed by the query specified defined in the body of the clause for *key*.

The goal of this paper is to explore how such typical ILP problems can be encoded in ASP languages. So, we will need to translate the typical ILP or Prolog construction into an ASP format. In the present paper, we employ IDP, which belongs to the ASP family of formalisms. Most statements and constraints written in IDP can be translated into standard ASP mechanically. A brief example-based introduction to IDP is in Appendix A and for a detailed system and paradigm description we refer to the IDP system and language description [2] and to the ASP primer [23].

To realize frequent query mining in IDP, we need to tackle four problems: (1) encode the queries in IDP; (2) implement the subsumption test to check whether a query subsumes a particular entity in the database; (3) choose and encode a language bias; and (4) determine, in addition to frequency, further constraints that could be used and encode them. We will now address each of these problems in turn.

3   Encoding

We assume that the dataset D is encoded as two predicates $edge(g, e_1, e_2)$, described before, and the ternary relation $label(g, n, l)$ that states that there is a node n with label l in graph g (for a discussion on how to extend the approach, see Sect. 6).

Encoding a Query. The coverage test in ILP is often θ- or OI-subsumption.

Definition 1 (OI and θ-subsumption [18]). *A clause c_1 θ-subsumes a clause c_2 iff there exists a substitution θ such that $c_1\theta \subseteq c_2$. Furthermore, c_1 OI-subsumes c_2 iff there exists substitution θ such that $com(c_1)\theta \subseteq com(c_2)$, where $com(c) = c \cup \{t_i \neq t_j \mid t_i \text{ and } t_j \text{ are two distinct terms in } c\}$.*

As ASP and, in particular, IDP are model-generation approaches, they always generate a set of ground facts. This implies that ultimately the queries will have to be encoded by a set of ground facts for *edge* and *label* as well (e.g., $\{edge(q, 77, 78),\ edge(q, 77, 79),\ label(q, 77, a),\ label(q, 78, b),\ label(q, 79, c), \dots\}$) and that we need to explicitly encode subsumption testing, rather than, as in Prolog, simply evaluate the query on the knowledge base. We use the convention that if a quantifier for a variable is omitted, then it is universally quantified. Furthermore, we use the convention that variables start with an upper-case and constants with a lower-case character. To illustrate the idea, consider the program in Eq. 1 that has a model if and only if the query q *OI*-subsumes the graph g. If we remove the last constraint in Eq. 1, we obtain θ-subsumption. Notice that the function θ will be explicit in the model. This program can be executed in IDP directly.

$$
\begin{aligned}
edge(q, X, Y) &\implies edge(g, \theta(X), \theta(Y)). \\
label(q, X, L) &\implies label(g, \theta(X), L). \\
X \neq Y &\implies \theta(X) \neq \theta(Y).
\end{aligned}
\tag{1}
$$

Notice that, in theory, it would be possible to simply encode subsumption testing in IDP or ASP as rule evaluation, that is, in the above example, to assert the knowledge base and to define the rule *key* and then to ask whether *key*(g) succeeds. Computationally, this would however be infeasible as the size of the grounded rules grows exponentially with the number of distinct variables in the rule.

Throughout the paper we use OI-subsumption for all tasks, except of the experimental comparison with Subsumer in Sect. 5 (since, it is not designed to perform OI-subsumption).

Encoding the Language Bias. In practice, one often bounds query languages, for instance by using a bottom clause and only considering queries or clauses that subsume the bottom clause.

Definition 2 (Language bias of a bottom clause $\perp$). *Let $\perp$ be a clause that we call bottom, then the language bias L is a set of clauses: $L = \{c \mid c$ OI-subsumes $\perp\}$.*

This approach also works for ASP. We select one graph from the data and the set of all atoms in that instance will serve as bottom clause $\perp$. When fixing

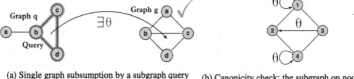

(a) Single graph subsumption by a subgraph query   (b) Canonicity check: the subgraph on nodes 1-3-4 is
             ($inq(x)$ in red)   mapped to 1-2-4 (lexicographically smaller)

Fig. 1. Examples of single graph subsumption (left) and of a canonicity check (right) (Colour figure online)

such an instance, we can encode queries by listing the nodes in that entity that will be present in the pattern using the unary predicate $inq(x)$ (for *in query*), as visualized in Fig. 1a.

We now present a modification of the previous encoding in Eq. 1 that takes into account the selection of the subgraph of the picked graph q as a query (marked in red in Fig. 1a). We refer to the edges in the bottom clause as *bedge* and labels in the bottom clause as *blabel*. When we refer to a node in the bottom clause, we call it a *bnode*.

$$inq(X) \wedge inq(Y) \wedge bedge(X,Y) \implies edge(g, \theta(X), \theta(Y)).$$
$$inq(X) \wedge blabel(X,L) \implies label(g, \theta(X), L). \qquad (2)$$
$$inq(X) \wedge inq(Y) \wedge X \neq Y \implies \theta(X) \neq \theta(Y).$$

The intuition behind these rules is that we select a subgraph by picking nodes in the bottom clause (i.e., a graph), and then we enforce the constraints on the nodes that have been selected. Equation 2 implements this by adding an *inq* predicate at the beginning of each clause as a guard, the rule is activated iff the corresponding node is selected.

Encoding the Multiple Subsumption Test. In frequent query mining one is interested in mining *frequent* queries, which implies that there is a bag of graphs to match.

Figure 2a illustrates this setup. We can see that we need to test whether the query θ-subsumes each of the graphs in the dataset. To do so, we quantify over a function representing θ, the homomorphism. This makes this formulation second order. Why do we quantify over a function here and not in the previous example? Before, the function was quantified existentially since the whole program was asking for a model, which is the same as asking whether the function exists. Here, however, we need a separate function for each graph, since the reasoning process is going to take into account the existence and non-existence of homomorphisms for particular graphs and to reason on top of that.

$$homo(G) \iff \exists \theta : \big(bedge(X,Y) \wedge inq(X) \wedge inq(Y) \implies edge(G, \theta(X), \theta(Y)).$$
$$blabel(X,L) \implies label(G, \theta(X), L).$$
$$X \neq Y \implies \theta(X) \neq \theta(Y)\big).$$

$$(3)$$

This constraint mimics the single graph subsumption test but introduces a new predicate *homo*, indicating a matched graph.

We refer to the group of constraints in Eq. 3 as *Matching-Constraint*. The syntax used above is not yet supported by ASP-solvers such as clasp [4] or IDP [2], but we will argue that adding such second order logic syntax is crucial to enable effective modeling and solving of any structural mining problem. This argument will be backed up by experiments in Sect. 5.

Encoding the Frequency Constraint. We consider two typical query mining settings: frequent and discriminative query mining. In the frequent setting a query is accepted if it subsumes at least t graphs. In the discriminative setting, each graph in the dataset is labeled as either positive or negative, as indicated by the corresponding predicates $positive(G)$ ($negative(G)$) that marks a positive (negative) graph G, and we are interested in queries that match more than t_p graphs with a positive label and do not match more than t_n negatively labeled graphs.

To model the frequent setting we use an aggregation constraint (Eq. 4) and the discriminative setting can be modeled similarly (Eq. 5):

$$|\{G : homo(G)\}| \geq t. \tag{4}$$

$$|\{G : positive(G) \wedge homo(G)\}| \geq t_p \wedge |\{G : negative(G) \wedge homo(G)\}| \leq t_n. \tag{5}$$

Encoding the Canonical form Constraint. It is well-known since the seminal work of Plotkin [17] that the subsumption lattice contains many clauses that are equivalent, and there has been a lot of work in ILP devoted to avoid the generation of multiple clauses from the same equivalence class (e.g., the work on optimality of refinement operators [15] and many others [12,13]).

This can be realized in ASP by checking that the current query is not isomorphic to a lexicographically smaller subgraph (using a lexicographic order on the identifiers of the entities or nodes). For example, in Fig. 1b, consider the subgraph induced by the nodes 1-3-4 (the numbers are identifiers and the colors are labels). Notice that there is an isomorphic subgraph 1-2-4, i.e., there exists a function θ preserving edges and labels such that a string representation of the latter subgraph is smaller than the former. We call the graph with the smallest lexicographic representation *canonical*. Canonicity can be enforced by the following group of constraints, called *Canonical-Form-Constraint*.

$$\begin{aligned}
&\neg\exists\theta\big(X \neq Y \implies \theta(X) \neq \theta(Y).\\
&\quad inq(X) \iff \exists Y : \theta(X) = Y.\\
&\quad inq(X) \wedge inq(Y) \wedge bedge(X,Y) \iff bedge(\theta(X),\theta(Y)).\\
&\quad inq(X) \wedge blabel(X) = Y \iff blabel(\theta(X)) = Y.\\
&\quad in\theta(Y) \iff \theta(X) = Y.\\
&\quad d_1(X) \iff inq(X) \wedge \neg in\theta(X).\\
&\quad d_2(X) \iff in\theta(X) \wedge \neg inq(X).\\
&\quad min(d_1(X)) > min(d_2(X))).
\end{aligned} \tag{6}$$

We refer to this group of constraints as *Canonical-Form-Constraint*. It enforces a query to have the smallest lexicographic representation, like the canonical code of mining algorithms [16,19]. We define a canonical representation in terms of the lexicographic order over the bottom clause node identifiers. If there is a query satisfying the constraints, then there is no other lexicographically smaller isomorphic graph.

Note: The intuition of Eq. 6 that the first four rules ensure the existence of a homomorphism under OI-assumption (which can be relaxed by removing a constraint) between the current query on the node in inq and another subgraph of the bottom clause. The other rules ensure that the other subgraph has a smaller lexicographic order: $in\theta$ is an auxiliary predicate that stores nodes of the other subgraph, in d_1 (d_2) there are nodes that only belong to the first, current query, (or second) subgraph. If the minimal node in d_1 (current query) is larger than in d_2 (the other graph), the query is not in canonical form.

We now present additional types of constraints that allow solving variations of the query mining problem.

Connectedness-Constraint. As in graph mining [16], we are often only interested in connected queries. The constraint to achieve this consists of two parts: 1) a definition of path and 2) a constraint over path. The first is defined inductively over the bnodes selected by $inq(X)$, the second enforces that any two nodes in the pattern are connected. Note, that the usage of $\{\dots\}$ indicates an inductive definition: the predicate on the left side is completely defined by the rules specified between curly brackets as a transitive closure. If both variables are in the pattern, there must be a path between them.

$$\{path(X,Y) \leftarrow inq(X) \wedge inq(Y) \wedge bedge(X,Y).$$
$$path(X,Y) \leftarrow \exists Z : inq(Z) \wedge path(X,Z) \wedge bedge(Z,Y) \wedge inq(Y).$$
$$path(Y,X) \leftarrow path(X,Y).\}$$
$$inq(X) \wedge inq(Y) \wedge X \neq Y \implies path(X,Y). \tag{7}$$

The Objective-Function. An objective function is one way to impose a ranking on the queries. A constraint in the form of an objective function is defined over a model to maximize certain parameters. We consider only objective functions over the queries and matched graphs in the dataset: (1) no objective function, i.e., the frequent query mining problem; (2) maximal size of a query Eq. 8 (in terms of bnodes) (3) maximal coverage Eq. 9 (i.e., the number of matched graphs); (4) discriminative coverage Eq. 10, i.e., the difference between the number of positively and negatively labeled graphs that are covered by a query.

$$|\{X : inq(X)\}| \mapsto \max \tag{8}$$
$$|\{G : homo(G)\}| \mapsto \max \tag{9}$$
$$|\{G : positive(G) \wedge homo(G)\}| - |\{G : negative(G) \wedge homo(G)\}| \mapsto \max \tag{10}$$

Mining the top-k queries with respect to a given objective function is often a more meaningful task than enumerating all frequent queries, since it provides a more manageable number of solutions that are best according to some function [24].

Topological-Constraint. Enforces parts of the bottom clause to be in the query. Let $\mathcal{X}$ be the desired subset of the nodes, then the constraint is $\bigwedge_{X \in \mathcal{X}} inq(X)$.

Cardinality-Constraint. This constraint ensures that the size of a graph pattern is at least n (at most, equal) $\exists \circ n \ X : inq(X)$, where $\circ \in \{=, \leq, <, \geq, >\}$.

If-Then-Constraint. This constraint ensures that if a node is present in the query, then another node must be present in the query, e.g., a node Y must be present in the query, if a node X is in. We encode this constraint as a logical implication: $inq(X) \implies inq(Y)$.

4   First Order Model

In the previous section we have given a positive answer to Question $\mathbf{Q_1}$ using second order logic. However, after formalizing a problem the next step is to actually solve it. In this section we address Question $\mathbf{Q_2}$: can we use existing ASP engines to solve these frequent query mining problems directly? Is it necessary to add constructs to existing modeling languages, or is it possible to write down an efficient and elegant first order logic model for which existing solvers can be used?

To answer Question $\mathbf{Q_2}$, we will encode the problem of frequent query mining as an ASP problem using enumeration techniques. We shall also show how to approach top-k querying mining with this encoding.

The Matching and Occurrence Constraints. Since we are restricted to FOL here, we have to encode θ as a binary predicate now, adding graph G as a parameter. In the encoding there are five clauses: the first enforces edge preservation; the second enforces that a mapping exists only for the bnodes in the pattern and only for the matched graphs, i.e., $\theta(G, X)$ is a partial function; the third enforces $\theta(G, X)$ to be an injective function on the bnodes for each of the graphs; the fourth enforces label matching; the fifth ensures occurrence frequency (the same as before).

$$homo(G) \wedge inq(X) \wedge inq(Y) \wedge bedge(X, Y) \implies edge(G, \theta(G, X), \theta(G, Y)).$$
$$homo(G) \wedge inq(X) \iff \exists Y : Y = \theta(G, X).$$
$$homo(G) \wedge inq(X) \wedge inq(Y) \wedge X \neq Y \implies \theta(G, X) \neq \theta(G, Y). \qquad (11)$$
$$homo(G) \wedge inq(X) \wedge blabel(X, L) \implies label(G, \theta(G, X), L).$$
$$|\{G : homo(G)\}| \geq t.$$

Encoding the Model Enumeration Constraints. A common technique in existing clause learning solvers with restarts for generating all solutions, is by asserting

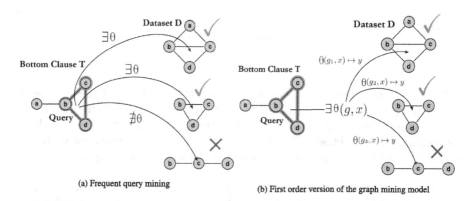

(a) Frequent query mining
(b) First order version of the graph mining model

Fig. 2. Conceptual comparison between first and second order models

for each found solution a model invalidating clause which excludes this model. We use it to denote that certain combinations of nodes in the bottom clause are no longer valid solutions. E.g., if we find a graph on nodes 1-2-3 to be a solution and we store these nodes, then we prohibit this combination of nodes to ensure that the solver will find a new different solution.

We present a version of *model invalidation clauses (MIC)* [1] for frequent query enumeration. Our *MIC*s are designed for Algorithm 1, since we enumerate queries of a fixed length: once we increase the length we remove all previous *MIC*s. This allows us to use *MIC*s of a very simple form: given a set of nodes $\mathcal{X}$, we define a constraint $C_\mathcal{X}$ as $C_\mathcal{X} = \bigwedge_{X \in \mathcal{X}} inq(X)$. For example, if a query q on the nodes 1-2-3 is found to be frequent, we would generate the following constraint, as so not to generate 1-2-3 again: $inq(1) \wedge inq(2) \wedge inq(3)$. Note, that this simple form of model invalidation clauses can only be used if an algorithm iterates over the length of a query and removes *MIC*s once the length is increased.

Algorithm 1. First Order Model: Iterative Query Enumeration

Input: D, k, t ▷ Dataset, #Queries, Threshold
Output: *queries* – set of frequent queries
queries ← ∅
⊥ ← *pick-language-bias(D)*   ▷ Language bias obtained from data
maxsize ← #*nodes*(⊥)
$i \leftarrow 0$
for *size* ∈ 1 ... *maxsize* **do**
    MICs ← ∅
    while True **do**
        query ← *mine-query(D, ⊥, t, size, MICs)*   ▷ IDP call, Eq. 11
        if *query* is None **then**
            break   ▷ the line below: IDP call, adopted Eq. 6
        canonical, isomorphic ← *get-canonical-and-isomorphic(query, ⊥)*
        queries ← *queries* ∪ {*canonical*}
        MICs ← *MICs* ∪ *isomorphic*
        $i \leftarrow i + 1$
        if $i \geq k$ **then return** *queries*
    return *queries*

Anti-monotonicity Property. It is easy to integrate avoidance of the infrequent supergraph generation in Algorithm 1: once a query is established to be infrequent, the corresponding *MIC* is kept, when the query length is increased. However, our experiments in Sect. 5 indicate that the current computational problems come from a different source and cannot be solved using this property.

Frequent Query Enumeration in Algorithm 1. We enumerate all frequent queries starting from the smallest ones (with only 1 node) to the largest ones (the bottom clause). Algorithm 1 has two loops. The first sets the current query size and sets

MIC to be empty (since we do not want to prohibit generation of supersets of already found queries). In the inner loop, we obtain a candidate for a frequent query by calling IDP once, then we check if the query is in canonical form and also obtain all isomorphic queries to this canonical form. After that we generate a *MIC* for each of them and prohibit the whole isomorphic class of queries to be generated. Note that generating all isomorphic queries is prohibitive, that is why we obtain a canonical query and remove all other isomorphic queries. The algorithm terminates if either it cannot find a new frequent query of any size or the required number of queries has been enumerated.

Top-k Problem. Current ASP solvers, including IDP, can perform optimization described in Constraints 8, 9 and 10. However, Algorithm 1 enumerates patterns with respect to their size and therefore needs to be modified. The key change is to remove the outer *for* loop with the size variable together with *Cardinality-Constraint*. In the experiment section we demonstrate that even top-1 is already excessively complex for modern ASP solvers and requires further investigation and development of the systems.

5 Experiments

In this section we evaluate the encoding on three problems: (1) classical θ-subsumption performed by IDP as encoded in Eq. 1, (2) the first order model in Algorithm 1 on the frequent query mining task, and (3) the first order model on the top-1 query mining task. In all experiments the frequency threshold t is set to 5 %. Since the task involves making a stochastic decision, i.e., *pick-language-bias* in Algorithm 1 picks a graph from the dataset at random, this choice may significantly influence the running time. To resolve this issue, for the graph enumeration problem we average over multiple runs for each dataset: each run involves the enumeration of many queries, i.e., each run generates many data points (runtimes to enumerate N queries). For the top-one mining problem, we present multiple runs for each dataset, since each run computes only one data point (runtime for the top-1 query). All experiments have been executed in a single thread on a 64-bit Ubuntu machine with Intel Core i5-3570 CPU @ 3.40GHz x 4 and 8GB memory.

Subsumption. We evaluate how the IDP model 1 encoding of θ-subsumption compares with subsumption engine The Subsumer [22]. We used the data from the original Subsumer experiments (transition phase on the subsumption hardness [18, p.327]) and evaluated IDP subsumption programs and Subsumer on a single hypothesis-example test, i.e., for each hypothesis and example we have made a separate call to IDP and Subsumer to establish subsumption.

The goal of this experiment is to compare how both systems perform if computations are done on a single example. We would like to estimate the potential gain in IDP, if we could specify a homomorphism existence check for each graph independently, like in a higher-order model Eq. 3, i.e., for each graph we

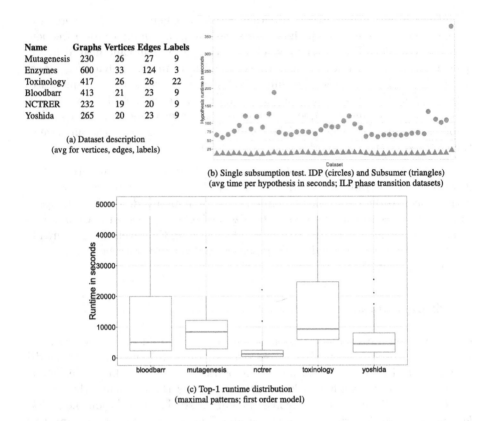

(a) Dataset description
(avg for vertices, edges, labels)

Name	Graphs	Vertices	Edges	Labels
Mutagenesis	230	26	27	9
Enzymes	600	33	124	3
Toxinology	417	26	26	22
Bloodbarr	413	21	23	9
NCTRER	232	19	20	9
Yoshida	265	20	23	9

(b) Single subsumption test. IDP (circles) and Subsumer (triangles)
(avg time per hypothesis in seconds; ILP phase transition datasets)

(c) Top-1 runtime distribution
(maximal patterns; first order model)

Fig. 3. Dataset description and the summary of subsumption and top-1 experiments

would check existence of some $\theta(X)$ instead of checking existence of one function $\theta(G, X)$ like in the first order model.

Figure 3b indicates that IDP and Subsumer perform within a constant bound when we make a separate call for each example and hypothesis. That is, for all but one dataset, the runtimes are within the same order of magnitude for the two methods. If the internal structures of the system are reused and the call is made only for one hypothesis per *set* of examples, we observe a speedup of at least one order of magnitude in Subsumer. This indicates that the system is able to efficiently use homomorphism independence. Once the solver community will have built extensions of systems like IDP to take advantage of homomorphism independence, the resulting systems will perform substantially better than the current ones and their performance will be close to that of special purpose systems such as Subsumer. More precisely, systems should exploit that it is computationally easier to find n functions $\theta(X)$, than one $\theta(G, X)$ for n values of G.

Datasets and Implementation. We now evaluate the query mining models on a number of well-known and publicly available datasets: Blood Barrier, NCTRER and Yoshida datasets are taken from [21], Mutagenesis and Enzymes datasets are from [11], Toxinology dataset is from [7]. A summary of the dataset properties is presented in Fig. 3a.

Top-1 Performance Evaluation. We present the results of evaluating the FOL model for top-1 mining in Fig. 3c. Results were only obtained for the maximal size, and no solutions were computed for Enzymes in a reasonable time (<10 h). The discriminative setting cannot be modeled and solved, since we need higher order primitives and for maximal coverage we already experienced an explosion of the search space. This experiment demonstrates that current satisfiability solvers cannot effectively perform search on both categories of query variables and homomorphism coverage; solver extensions are necessary. The first extension is to specify independence of the homomorphisms in the model, i.e., higher order primitives as indicated in the model Eq. 3, the second is to give more control over the solver's decisions, e.g., by allowing to specify a decision order.

Frequent Query Mining. The goal of the next experiment is to estimate the potential gain of the introduction of a higher order modeling construction. To do this, we mimicked the higher order behavior in Algorithm 1 by making several calls to IDP (one per graph) in the method *mine-query*. We call this mimicking model *decomposed*. The results of the evaluation are presented in Fig. 4a, a comparison of the decomposed and the original FOL model are in Fig. 4b. Homomorphism dependence in the first order model causes serious computational difficulties, which makes higher order primitives one of the main priorities for solving query mining. The runtime also grows with the size of a pattern since it affects the search space and runtime as a result.

Summary of the Experimental Results. The results show that the model performs reasonably and we therefore conclude that it can be considered a first step towards the development of declarative languages for relational query mining. From Figs. 3b and 4a it is clear that the core computational difficulty lies in the inability to state homomorphism independence. This follows from the candidate

dataset	model	1	5	10	25	50
mutagenesis	FOL	35s	2m25s	7m22s	23m22s	1h29m
	decomposed	17s	43s	84s	3m50s	9m26s
nctrer	FOL	30s	3m24s	7m17s	24m28s	1h27m
	decomposed	5s	35s	1m16s	6m0s	9m20s
yoshida	FOL	35s	2m53s	6m34s	31m13s	1h56m
	decomposed	9s	1m1s	2m0s	5m48s	15m8s
bloodbarr	FOL	1m4s	6m26s	15m51s	1h47m	8h2m
	decomposed	9s	57s	2m3s	9m55s	14m52s
toxinology	FOL	1m21s	8m20s	22m5s	2h42m	16h3m
	decomposed	11s	56s	2m35s	7m54s	18m5s
enzymes	FOL	7m3s	45m48s	4h0m	—	—
	decomposed	15s	1m27s	2m57s	13m26s	38m16s

(a) Frequent query enumeration time of N queries

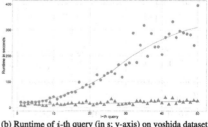

(b) Runtime of i-th query (in s; y-axis) on yoshida dataset (x-axis is the query index i)

Fig. 4. Frequent query enumeration FOL (blue) vs decomposed model (red) (Color figure online)

generation and canonicity check runtime and from the speedup that Subsumer has when it is applied to the whole set of examples with one call. After comparing the performance of IDP and Subsumer on a single example and observing the speed up of the decomposed model in Fig. 4a, one would expect a significant speedup if it were implemented within the solver.

We have observed in Fig. 3c that query mining introduces interesting computational challenges and might be of interest to the solver development community as well as to the ILP community. We also pointed out the reasons for these computational difficulties and suggested possible ways to enhance the solvers.

6   Model Discussion and Generalization

Advantages and Disadvantages of the Model. There is a number of advantages of the declarative approach to pattern mining as compared to the classical imperative methods: compact and clear representation; extendability and generality; provability and formal semantics of the models; reliable and portable implementation (the solvers developed by the community, well tested and already applied to a variety of tasks; the solvers are available for all popular operation systems: Linux, MacOS and Windows).

In particular, our model represents compactly not only the model of graph mining problem but also the code that can be executed to solve the problem. Adding the source code of gSpan [25] to a paper (~2000 lines of C++) would be impossible. Also the model can be easily extended by adding constraints to the theory, e.g. adding a constraint to handle labels on edges is just an extra line with a straightforward logical formula. The last but not least our formulation allows formal rigorous reasoning on the constraints that constitute the model.

These advantages come, of course, not without a cost. Typically, specialized algorithms perform an order of magnitude (at least) faster. The main reason for such a speed up is the heuristic approach incorporated in most of the mining algorithms that takes advantage of the structure of the problem. Our experiments in Fig. 4b demonstrate that the declarative approach can reduce the runtime gap by extending the language to better incorporate the structure of a problem. Also, modeling this kind of mining tasks influences the way declarative solvers are built and potentially can lead to better solving systems that would perform reasonable on mining takes.

Generalization of the Approach. Let us demonstrate how the model can be extended to handle labels on the edges as an example how declarative systems can be adapted to solve new tasks.

Assume that the edge labels are represented using the predicate $edgelabel(g, e_1, e_2, l)$ and the edge labels of the bottom clause are stored in the predicate $tedgelabel(e_1, e_2, l)$. Then, the first order formulation Eq. 11 can be extended by adding the constraint:

$$inq(X) \wedge inq(Y) \wedge tedgelabel(X, Y, L) \implies edgelabel(G, \theta(G, X), \theta(G, Y), L).$$

Intuitively, this constraint ensures that if there is an edge (X, Y) with a label L in the bottom clause, then there is an edge with a l label L in the graph G. If each edge has a label, the constraint above can replace the first constraint in Eq. 11.

The smallest change, according to the principle, is the addition of a rule or a fact.n general this example demonstrates how declarative approach and our model in particular implements the elaboration tolerance principle [14], i.e. a small change in the problem formulation, should introduce a small change in the model. The smallest changes, according to the principle, is the addition of a rule or a fact. With this respect our model satisfies elaboration tolerance principle and can be called an additive elaboration.

7 Related Work

WARMR [10] and FARMER [16] are extensions of the Apriori algorithm for discovering frequent structures in multiple relations. Even though they use ILP techniques (e.g., a declarative language bias) to determine frequent queries, they imperatively specify computations and the algorithms do not support the addition of arbitrary constraints due to the restricted nature of the algorithm, and hence focus on a more specific task. C-Farmr [19] is an ILP system for frequent Datalog clauses that uses so-called condensed representations to avoid the generation of semantically equivalent clauses.

The XHAIL system [20] uses ASP as a computational engine to perform abductive inductive reasoning. It is similar in the way we use an ASP engine as computational core, but XHAIL focuses on the abduction task, whereas we focus on query mining which always involves aggregation and model enumeration.

In the work on sequence testing [3] ASP is used as a subroutine in a cycle and model projection on a predicate is also present similarly to our Algorithm 1, but this similarity is technical, since the tasks are of a different nature. ASP has also been applied to itemset mining [9] and sequence mining [6], but these methods only deal with one particular task in its basic formulation.

gSpan [25] and B-AGM, Biased-Apriori-based Graph Mining [8], are specialized algorithm designed to solve the frequent graph pattern mining problem. The algorithm is tailored to solve only one task exclusively and requires significant changes in its core to be extended to solve other mining tasks (and, to the best of our knowledge, there is no extension to the full relation setting). It is similar, however, in the computational challenges: canonicity checks, homomorphism checks, language bias etc.

8 Conclusions

We have shown that modern ASP solvers, in principle, can be applied to ILP query mining tasks. We have provided experimental evidence that these models can be used as prototypes for developing declarative mining languages. We have also indicated the reasons why the solvers could be extended to make computation efficient and proposed concrete extensions as well as estimated their potential effectiveness. The query mining models and the experimental setup we developed provide an interesting challenge for the ASP community and a potentially useful tool for the ILP community.

A   Appendix: Introduction to IDP

Listing 1.1. IDP source code example – map coloring

```
vocabulary V{
  type area
  type color
  border(area,area)
  coloring(area):color
}
theory T:V{
  // Adjacent countries can not have the same color
    ∀A₁A₂ : border(A₁, A₂) ⟹ coloring(A₁) ≠ coloring(A₂).
}
structure S:V{
  area={belgium; holland; germany; luxembourg; austria; swiss;france }
  color={blue;red;yellow;green}
  border={
   (belgium,holland);(belgium, germany);(belgium,luxembourg);(belgium,france);
   (holland,germany);(germany,luxembourg);(germany,austria);(germany,swiss);
   (germany,france);(luxembourg,france);(austria,swiss);(swiss,france)
  }
}
```

The IDP language [1,2] is an extension of first order logic with inductive definitions and aggregration. The IDP system implements finite satisfiability and can be considered as an ASP system.

The particular type of inference we use in our work is *model expansion*. The task of model expansion is to expand a finite interpretation S for the subvocabulary V of a given logic theory T to a model of T. In the example above, V is the vocabulary of the map colouring problem i.e. *area, color, border(area, area)*, *coloring :: area ↦ color*, S consists of 7 countries, 4 colours and a border relation between the countries; T is the constraint that two bordering countries cannot have the same color.

In this example, the model expansion task is to find an extension of S, i.e. *coloring* function, such that the constraint in T is satisfied, i.e. all bordering countries have different colours.

The example can be tried online (select file "Map Colouring"):

adams.cs.kuleuven.be/idp/server.html.

References

1. De Cat, B.: Separating Knowledge from Computation: An FO(.) Knowledge Base System and its Model Expansion Inference. Ph.D. thesis. KU Leuven
2. De Cat, B., Bogaerts, B., Bruynooghe, M., Denecker, M.: Predicate Logic as a Modelling Language: The IDP System. In: CoRR abs/1401.6312 (2014)

3. Erdem, E., Inoue, K., Oetsch, J., Puehrer, J., Tompits, H., Yilmaz, C.: Answer-set programming as a new approach to event-sequence testing. In: International Conference on Advances in System Testing and Validation Lifecycle (2011)
4. Gebser, M., Kaufmann, B., Neumann, A., Schaub, T.: *clasp*: a conflict-driven answer set solver. In: Baral, C., Brewka, G., Schlipf, J. (eds.) LPNMR 2007. LNCS (LNAI), vol. 4483, pp. 260–265. Springer, Heidelberg (2007)
5. Guns, T., Dries, A., Tack, G., Nijssen, S., De Raedt, L.: MiningZinc: a modeling language for constraint-based mining. In: IJCAI (2013)
6. Guyet, T., Moinard, Y., Quiniou, R.: Using Answer Set Programming forpattern mining. In: CoRR abs/1409.7777 (2014)
7. Helma, C., Kramer, S.: A survey of the predictive toxicology challenge 2000–2001. Bioinformatics **19**(10), 1179–1182 (2003)
8. Inokuchi, A., Washio, T., Motoda, H.: An apriori-based algorithm for mining frequent substructures from graph data. In: Zighed, D.A., Komorowski, J., Żytkow, J.M. (eds.) PKDD 2000. LNCS (LNAI), vol. 1910, pp. 13–23. Springer, Heidelberg (2000)
9. Järvisalo, M.: Itemset mining as a challenge application for answer set enumeration. In: Delgrande, J.P., Faber, W. (eds.) LPNMR 2011. LNCS, vol. 6645, pp. 304–310. Springer, Heidelberg (2011)
10. King, R.D., Srinivasan, A., Dehaspe, L.: Warmr: a data mining tool for chemical data. J. Comput. Aided Mol. Des. **15**(2), 173–181 (2001)
11. Kumar, D.A., de Compadre, L., Rosa, L., Gargi, D., Shusterman Alan, J., Corwin, H.: Structure-activity relationship of mutagenic aromatic and heteroaromatic nitro compounds. Correlation with molecular orbital energies and hydrophobicity. J. Med. Chem. **34**(2), 786–797 (1991)
12. van der Laag, P.R.J., Nienhuys-Cheng, S.-H.: Completeness and properness of refinement operators in inductive logic programming. J. Log. Program. **34**(3), 201–225 (1998)
13. Lehmann, J., Hitzler, P.: Foundations of refinement operators for description logics. In: Blockeel, H., Ramon, J., Shavlik, J., Tadepalli, P. (eds.) ILP 2007. LNCS (LNAI), vol. 4894, pp. 161–174. Springer, Heidelberg (2008)
14. McCarthy, J.: Elaboration Tolerance (1999)
15. Muggleton, S., Entailment, I.: Inverse entailment and progol. New Gener. Comput. Spec. Issue Inductive Logic Program. **13**(3–4), 245–286 (1995)
16. Nijssen, S., Kok, J.N.: Efficient frequent query discovery in FARMER. In: Lavrač, N., Gamberger, D., Todorovski, L., Blockeel, H. (eds.) PKDD 2003. LNCS (LNAI), vol. 2838, pp. 350–362. Springer, Heidelberg (2003)
17. Plotkin, G.D.: A further note on inductive generalization. In: Machine Intelligence, vol. 6, pp. 101–124 (1971)
18. De Raedt, L.: Logical and Relational Learning: From ILP to MRDM (Cognitive Technologies). Springer-Verlag New York Inc., New York (2008)
19. De Raedt, L., Ramon, J.: Condensed representations for inductive logic programming. In: KR, pp. 438–446 (2004)
20. Ray, O.: Nonmonotonic abductive inductive learning. J. Appl. Logic **3**, 329–340 (2008)
21. Rückert, U., Kramer, S.: Optimizing feature sets for structured data. In: Kok, J.N., Koronacki, J., Lopez de Mantaras, R., Matwin, S., Mladenič, D., Skowron, A. (eds.) ECML 2007. LNCS (LNAI), vol. 4701, pp. 716–723. Springer, Heidelberg (2007)
22. Santos, J., Muggleton, S.: Subsumer: A Prolog theta-subsumption engine. In: ICLP Technical Communications, vol. 7, pp. 172–181 (2010)

23. Eiter, T., Ianni, G., Krennwallner, T.: Answer set programming: a primer. In: Tessaris, S., Franconi, E., Eiter, T., Gutierrez, C., Handschuh, S., Rousset, M.-C., Schmidt, R.A. (eds.) Reasoning Web. LNCS, vol. 5689, pp. 40–110. Springer, Heidelberg (2009)
24. Vreeken, J., Leeuwen, M., Siebes, A.: Krimp: mining itemsets that compress. Data Min. Knowl. Discov. **23**(1), 169–214 (2011)
25. Yan, X., Han, J.: gSpan: graph-based substructure pattern mining. In: ICDM (2002)

Learning Inference by Induction

Chiaki Sakama[1]([✉]), Tony Ribeiro[2], and Katsumi Inoue[3]

[1] Wakayama University, Wakayama, Japan
sakama@sys.wakayama-u.ac.jp
[2] Institut de Recherche en Communications et Cybernetique de Nantes,
Nantes, France
tony.ribeiro@irccyn.ec-nantes.fr
[3] National Institute of Informatics, Tokyo, Japan
inoue@nii.ac.jp

Abstract. This paper studies *learning inference* by induction. We first consider the problem of learning logical inference rules. Given a set S of propositional formulas and their logical consequences T, the goal is to find deductive inference rules that produce T from S. We show that an induction algorithm **LF1T**, which learns logic programs from interpretation transitions, successfully produces deductive inference rules from input transitions. Next we consider the problem of learning non-logical inference rules. We address three case studies for learning abductive inference, frame axioms and conversational implicature by induction. The current study provides a preliminary approach to the problem of learning inference to which little attention has been paid in machine learning and ILP.

1 Introduction

Induction has been used for learning regularities or general rules hidden in data sets. Hypotheses generated by induction are used not only for explaining given evidences but for predicting new phenomena. Induction in these tasks realizes *learning knowledge*—knowing that something is true. Concept learning and data mining are of this type. On the other hand, little attention has been paid for *learning inference*—knowing the ways of thinking by humans. Learning inference is the problem of finding inference rules or thinking patterns by humans. We make a variety of inferences in our daily life. Those inferences are classified into two categories: logical inference and non-logical inference. Logical inference is used in classical or non-classical logics. Non-logical inference includes empirical laws or pragmatic inference. The primary interest of this paper is to argue the possibility of developing machine learning algorithms that can automatically acquire rules of inferences.

To learn logical inference rules, Sakama and Inoue [16] introduce a conceptual framework of *learning logics*. In this framework, they introduce a method of learning propositional logic inference rules inductively from input data that consist of formulas and their logical consequences. In the current paper, we implement learning logical inference rules using the **LF1T** induction algorithm [5,14]

K. Inoue et al. (Eds.): ILP 2015, LNAI 9575, pp. 183–199, 2016.
DOI: 10.1007/978-3-319-40566-7_13

and examine whether it successfully produces inference rules of propositional natural deduction. Next we apply the method to learning non-logical inference rules. We address three case studies, learning rules for abduction, frame axioms, and conversational implicature. We discuss related issues and finally remark open questions. The rest of this paper is organized as follows. Section 2 provides a framework for learning logics and its implementation by **LF1T**. Section 3 addresses three cases of learning non-logical inference rules. Section 4 discusses related issues and Sect. 5 addresses final remarks.

2 Learning Logical Inference

2.1 Learning Logics

We review a conceptual framework for learning logics that is introduced in [16]. There are an agent $\mathcal{A}$ and a machine $\mathcal{M}$. The agent $\mathcal{A}$, which could be a human or a computer, is capable of deductive reasoning: it has a set $\mathcal{L}$ of axioms and inference rules in classical logic. Given a (finite) set S of formulas as an input, the agent $\mathcal{A}$ produces a (finite) set of formulas T such that $T \subseteq Th(S)$ where $Th(S)$ is the set of logical consequences of S. On the other hand, the machine $\mathcal{M}$ has no axiomatic system for deduction, while it is equipped with a machine learning algorithm $\mathcal{C}$. Given input-output pairs $(S_1, T_1), \ldots, (S_i, T_i), \ldots$ (where $T_i \subseteq Th(S_i)$) of $\mathcal{A}$ as an input to $\mathcal{M}$, the problem is whether one can develop an algorithm $\mathcal{C}$ which successfully produces an axiomatic system $\mathcal{K}$ for deduction. An algorithm $\mathcal{C}$ is *sound* wrt $\mathcal{L}$ if it produces an axiomatic system $\mathcal{K}$ such that $\mathcal{L} \vdash \mathcal{K}$. An algorithm $\mathcal{C}$ is *complete* wrt $\mathcal{L}$ if it produces an axiomatic system $\mathcal{K}$ such that $\mathcal{K} \vdash \mathcal{L}$ where $\vdash$ means the logical consequence under $\mathcal{L}$. Designing a sound and complete algorithm $\mathcal{C}$ is called a problem of *learning logics* (Fig. 1). In this framework, an agent $\mathcal{A}$ plays the role of a teacher who provides training examples representing premises along with entailed consequences. The output $\mathcal{K}$ is refined by incrementally providing examples. Figure 1 describes a deduction system $\mathcal{L}$ while it could be a system of arbitrary logic, e.g. nonmonotonic logic, modal logic, fuzzy logic, as far as it has a formal system of inference.

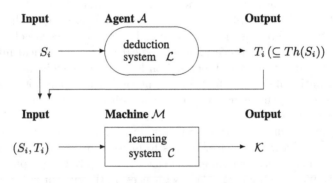

Fig. 1. Learning logics [16]

Sakama and Inoue [16] provide a simple case study for learning deduction rules of propositional logic. They represent a formal system $\mathcal{L}$ of propositional logic using *metalogic programming* and show that deductive inference rules can be reproduced as meta-rules of logic programs. In this paper, we implement learning deductive inference rules using the **LF1T** algorithm [5,14] and examine whether it successfully produces inference rules of propositional natural deduction.

2.2 Learning from 1-Step Transitions

Learning from 1-step transitions (**LF1T**) [5] is a framework for learning normal logic programs from transitions of interpretations. Here we apply **LF1T** for learning *definite logic programs*, a subclass of normal logic programs that do not contain negation as failure. Let $\mathcal{B}$ be the set of all ground atoms (Herbrand base) and P a (ground) definite logic program (or simply, a *program*) that consists of *rules* of the form:

$$a \leftarrow b_1, \ldots, b_n \quad (n \geq 0) \tag{1}$$

where $a, b_1, \ldots, b_n$ are ground atoms from $\mathcal{B}$. For each rule R of the form (1), the atom a is the *head* (written $h(R)$) and the conjunction $b_1, \ldots, b_n$ is the *body* of the rule (written $b(R)$). The body is identified with the set of atoms $\{b_1, \ldots, b_n\}$. A rule with the empty body is a *fact*. A rule with variables represents the set of ground instances. A rule R_1 *subsumes* another rule R_2 (written $R_1 \leq R_2$) if $h(R_1) = h(R_2)$ and $b(R_1) \subseteq b(R_2)$. For two programs P_i and P_j, define $P_i \sqsubseteq P_j$ iff for any $R \in P_i$ there is $R' \in P_j$ s.t. $R \leq R'$.

LF1T produces a program from a pair of interpretations as follows.

Input: $E \subseteq 2^{\mathcal{B}} \times 2^{\mathcal{B}}$
Output: a program P such that $J = T_P(I)$ holds for any $(I, J) \in E$

where $T_P(I) = \{ h(R) \mid R \in P \text{ and } b(R) \subseteq I \}$ [17]. A rule R is *consistent* with (I, J) if $b(R) \subseteq I$ implies $h(R) \in J$, otherwise, R is *inconsistent* with (I, J). A rule R is called an *anti-rule* with respect to a transition (I, J) if R is inconsistent with (I, J). A program P is *consistent* with (I, J) if every rule in P is consistent with (I, J). In **LF1T**, a positive example is input as a one-step state transition from I to J, which is given as a pair of Herbrand interpretations. **LF1T** outputs a single program that is consistent with all state transitions given in the input.

In this subsection, we use a *top-down* version of **LF1T** [14], which generates hypotheses by specialization from the most general rules until a program is consistent with all input state transitions. More precisely, **LF1T** starts with the initial program $P = \{ a \leftarrow \mid a \in \mathcal{B} \}$. Then **LF1T** iteratively analyzes each transition (I, J). For each atom a that does *not* appear in J, **LF1T** produces an anti-rule:

$$a \leftarrow \bigwedge_{b_i \in I} b_i . \tag{2}$$

The rule (2) is an anti-rule wrt (I, J) because $\bigwedge_{b_i \in I} b_i \subseteq I$ but $a \notin J$. Any rule of P that subsumes such an anti-rule is inconsistent with the transition and

---

Input : a set $\mathcal{B}$ of atoms and $E \subseteq 2^{\mathcal{B}} \times 2^{\mathcal{B}}$
Output : A definite logic program P such that $J = T_P(I)$ for any $(I, J) \in E$.

put $P := \{ a \leftarrow \mid a \in \mathcal{B} \}$;
while $E \neq \emptyset$ do
    pick $(I, J) \in E$; $E := E \setminus \{(I, J)\}$;
    for each $a \in \mathcal{B} \setminus J$, put
        $R_a^I := a \leftarrow \bigwedge_{b_i \in I} b_i$;
        for any rule $R \in P$ that subsumes R_a^I, put
            $\mathcal{R} := \{ h(R) \leftarrow b(R) \wedge c_j \mid c_j \in \mathcal{B} \setminus I \}$;
        Replace R with the rules of $\mathcal{R}$ and remove any rule $R' \in P$ that is subsumed by some
    rule in $\mathcal{R}$
return P

Fig. 2. Top-down **LF1T**[14]

must be revised. To this end, every rule R of P that subsumes (2) is *minimally specialized* by replacing R with the rules in $\{ h(R) \leftarrow b(R) \wedge c_j \mid c_j \in \mathcal{B} \setminus I \}$ to make P consistent with the new transition by avoiding the subsumption of all anti-rules produced by (I, J). After such minimal specialization, P becomes consistent with the new transition while remaining consistent with all previously analyzed transitions. The algorithm is sketched in Fig. 2. It is shown that **LF1T** produces a program P such that $J = T_P(I)$ for any $(I, J) \in E$ and minimal wrt the ordering $\sqsubseteq$ [14].[1]

    In the next section, we use the **LF1T** induction algorithm for learning deduction rules and show experimental results.

2.3   Learning Deduction Rules by LF1T

In this section, we use **LF1T** as a learning system $\mathcal{C}$ in Fig. 1. We assume a (propositional) *natural deduction system* $\mathcal{L}$ [13] represented by a *metalogic program* P. In P every propositional formula F is represented by a fact using a meta-predicate *hold* as

$$hold(F). \tag{3}$$

Using the expression, *Modus Ponens* is represented as a *meta-rule* in P as follows:

$$hold(G) \leftarrow hold(F \rightarrow G), hold(F) \tag{4}$$

where F and G are variables representing any propositional formula.

    Suppose that P contains the single rule of (4). Let p and q be propositional variables. Given the set $I = \{ hold(p), hold(p \rightarrow q) \}$, it becomes $T_P(I) = \{ hold(q) \}$. $T_P(I)$ represents a set of formulas that are deduced from I using Modus Ponens. In this way, a program P provides transitions (I, J)

---

[1] The result is shown for normal logic programs and is applied to their subclass of programs.

such that $J \subseteq T_P(I)$. Then, given (I, J) as input, our goal is to examine whether **LF1T** can reproduce correct inference rules of deduction represented by meta-rules in P. We use formulas represented by propositional variables (e.g. $p \rightarrow q$), rather than formulas represented by particular propositional constants (e.g. *human* $\rightarrow$ *animal*). With this setting, produced rules represent relations among formulas written by propositional variables, which are instantiated by any propositional constants. To learn inference rules, the Herbrand base $\mathcal{B}$ is firstly fixed as a set of facts of the form (3). Inputs are then constructed as pairs $(I, J) \in E$ where $E \subseteq 2^{\mathcal{B}} \times 2^{\mathcal{B}}$.

In what follows, we provide some results of experiments.

Let $\mathcal{B} = \{ hold(p), hold(q), hold(r), hold(p \rightarrow r) \}$. We address the process of constructing a rule with the atom $hold(r)$ in the head.

Step 0: LF1T starts with the most general rule:

$$hold(r) \leftarrow . \tag{5}$$

Step 1: Suppose that the transition $(\emptyset, \emptyset)$ is given. The rule (5) is inconsistent with this transition (because $b((5)) = \emptyset \subseteq \emptyset$ but $h((5)) = hold(r) \notin \emptyset$), so that (5) is an anti-rule wrt $(\emptyset, \emptyset)$ and is (minimally) specialized into four rules by introducing an atom from $\mathcal{B}$:

$$hold(r) \leftarrow hold(p) \tag{6}$$
$$hold(r) \leftarrow hold(q) \tag{7}$$
$$hold(r) \leftarrow hold(r) \tag{8}$$
$$hold(r) \leftarrow hold(p \rightarrow r). \tag{9}$$

Those rules are consistent with the transition $(\emptyset, \emptyset)$.

Step 2: Suppose that the transition $(\{hold(p)\}, \{hold(p)\})$ is given. The rule (6) is inconsistent with this transition, so that (6) is specialized into three rules:

$$hold(r) \leftarrow hold(p), hold(q)$$
$$hold(r) \leftarrow hold(p), hold(r)$$
$$hold(r) \leftarrow hold(p), hold(p \rightarrow r).$$

These three rules are respectively subsumed by the rules (7), (8) and (9) in Step 1, hence removed. As a result, the rules (7), (8) and (9) remain.

Step 3: Suppose that the transition $(\{hold(q)\}, \{hold(q)\})$ is given. The rule (7) is inconsistent with this transition, and is removed after specialization. As a result of subsumption, three rules (8), (9) and the newly constructed rule (10) remain.

$$hold(r) \leftarrow hold(p), hold(q). \tag{10}$$

Step 4: Suppose that the transition $(\{hold(p \rightarrow r)\}, \{hold(p \rightarrow r)\})$ is given. The rule (9) is inconsistent with this transition, and is removed after specialization. As a result of subsumption, two rules are newly constructed:

$$hold(r) \leftarrow hold(p \rightarrow r), hold(p) \tag{11}$$
$$hold(r) \leftarrow hold(p \rightarrow r), hold(q). \tag{12}$$

Now four rules (8), (10), (11) and (12) remain.

Step 5: Suppose that the transition $(\{hold(p), hold(q)\}, \{hold(p), hold(q)\})$ is given. The rule (10) is inconsistent with this transition, and is removed after specialization. As a result of subsumption, three rules (8), (11) and (12) remain.

Step 6: Suppose that the transition $(\{hold(p \to r), hold(q)\}, \{hold(p \to r), hold(q)\})$ is given. The rule (12) is inconsistent with this transition, and is removed after specialization. As a result of subsumption, two rules (8) and (11) remain.

Step 7: Suppose that the transition $(\{hold(p \to r), hold(p)\}, \{hold(p \to r), hold(p), hold(r)\})$ is given. Two rules (8) and (11) are consistent with this transition and remain as they are.

The remaining two rules (8) and (11) are consistent with any other transitions (I, J) such that J represents logical consequences of I under a metalogic program P. Then **LF1T** produces those rules as output. The rule (8) represents **Repetition (Rep)** and (11) represents **Modus Ponens (MP)**. The input-output of **LF1T** is summarized in Table 1.[2]

Table 1. LF1T input-output

input	output
$(\emptyset, \emptyset)$	$hold(r) \leftarrow hold(p).$
	$hold(r) \leftarrow hold(q).$
	$hold(r) \leftarrow hold(r).$
	$hold(r) \leftarrow hold(p \to r).$
$(\{hold(p)\}, \{hold(p)\})$	~~$hold(r) \leftarrow hold(p).$~~
	$hold(r) \leftarrow hold(q).$
	$hold(r) \leftarrow hold(r).$
	$hold(r) \leftarrow hold(p \to r).$
$(\{hold(q)\}, \{hold(q)\})$	~~$hold(r) \leftarrow hold(q).$~~
	$hold(r) \leftarrow hold(r).$
	$hold(r) \leftarrow hold(p \to r).$
	$hold(r) \leftarrow hold(p), hold(q).$
$(\{hold(p \to r)\}, \{hold(p \to r)\})$	$hold(r) \leftarrow hold(r).$
	~~$hold(r) \leftarrow hold(p \to r).$~~
	$hold(r) \leftarrow hold(p), hold(q).$
	$hold(r) \leftarrow hold(p \to r), hold(p).$
	$hold(r) \leftarrow hold(p \to r), hold(q).$
$(\{hold(p), hold(q)\}, \{hold(p), hold(q)\})$	$hold(r) \leftarrow hold(r).$
	~~$hold(r) \leftarrow hold(p), hold(q).$~~
	$hold(r) \leftarrow hold(p \to r), hold(p).$
	$hold(r) \leftarrow hold(p \to r), hold(q).$
$(\{hold(p \to r), hold(q)\}, \{hold(p \to r), hold(q)\})$	$hold(r) \leftarrow hold(r).$
	$hold(r) \leftarrow hold(p \to r), hold(p).$
	~~$hold(r) \leftarrow hold(p \to r), hold(q).$~~
$(\{hold(p \to r), hold(p)\}, \{hold(p \to r), hold(p), hold(r)\})$	$hold(r) \leftarrow hold(r).$
	$hold(r) \leftarrow hold(p \to r), hold(p).$

---

[2] The experimental archive is found at http://www.wakayama-u.ac.jp/~sakama/ILP2015-short/.

In the table, those inputs except the last one are of the form (I, I) that *does not* represent any change. Those examples are used for excluding rules describing incorrect transitions. By contrast, the last input is used for verifying whether the remaining rule represents correct transitions. Other results of experiments are addressed below.

- Given $\mathcal{B} = \{\, hold(p),\, hold(\neg p),\, hold(q),\, hold(\neg q),\, hold(p \to q),\, hold(q \to r),\, hold(p \to r)\,\}$, **LF1T** produces

$$hold(\neg p) \leftarrow hold(p \to q),\, hold(\neg q) \quad \textbf{(Modus Tollens (MT))}$$
$$hold(p \to r) \leftarrow hold(p \to q),\, hold(q \to r) \textbf{(Hypothetical Syllogism (HS))}$$

- Given $\mathcal{B} = \{\, hold(p),\, hold(\neg p),\, hold(q),\, hold(\neg q),\, hold(p \vee q),\, hold(\neg p \vee \neg q),\, hold(r \vee s),\, hold(\neg r \vee \neg s),\, hold(p \to r),\, hold(q \to s)\,\}$, **LF1T** produces

$$hold(p) \leftarrow hold(p \vee q),\, hold(\neg q) \quad \textbf{(Disjunctive Syllogism (DS))}$$
$$hold(r \vee s) \leftarrow hold(p \vee q),\, hold(p \to r),\, hold(q \to s)$$
$$\textbf{(Constructive Dilemma (CD))}$$
$$hold(\neg p \vee \neg q) \leftarrow hold(\neg r \vee \neg s),\, hold(p \to r),\, hold(q \to s)$$
$$\textbf{(Destructive Dilemma (DD))}$$

- Given $\mathcal{B} = \{\, hold(p),\, hold(q),\, hold(r),\, hold(p \wedge q),\, hold(q \wedge r),\, hold(p \vee q),\, hold(q \vee r)\,\}$, **LF1T** produces

$$hold(p) \leftarrow hold(p \wedge q) \quad \textbf{(Conjunction Elimination (CE))}$$
$$hold(p \wedge q) \leftarrow hold(p),\, hold(q) \quad \textbf{(Conjunction Introduction (CI))}$$
$$hold(p \vee q) \leftarrow hold(p) \quad \textbf{(Disjunction Introduction (DI))}$$

The number of examples used for learning above rules depends on $\mathcal{B}$. To obtain correct inference rules, a sufficient number of input examples are needed. To see whether a produced rule R is a correct inference rule, input test data I to R and check whether R produces a correct output J satisfying $T_P(I) = J$ for a metalogic program P.

In this way, **LF1T** successfully produces inference rules of natural deduction. Note that produced rules are applied to any propositions. For instance, in Modus Ponens

$$hold(q) \leftarrow hold(p \to q),\, hold(p) \tag{13}$$

p and q are propositional variables that can be instantiated by any propositional constants. Moreover, since any propositional formula is named by a new propositional variable, we can replace propositional variables p, q, r, ... with new propositional variables F, G, H, ... representing arbitrary propositional formulas. The rule (13) is then interpreted as a rule between formulas:

$$hold(G) \leftarrow hold(F \to G),\, hold(F).$$

As such, **LF1T** produces *inference schemata* of natural deduction. Let $\mathcal{L}$ be a natural deduction system that has ten inference rules—**Rep, MP, MT, HS, DS, CD, DD, CE, CI**, and **DI**. Then we have the next result.

Proposition 2.1 *Let P be a metalogic program representing a propositional natural deduction system $\mathcal{L}$. Then there is a finite number of inputs (I, J) satisfying $T_P(I) = J$ such that* **LF1T** *can reproduce the set of inference rules of $\mathcal{L}$ from them.*

Proof. As we have seen in this section, the ten rules of $\mathcal{L}$ can be obtained by **LF1T**. The natural deduction system $\mathcal{L}$ consists of inference rules represented by a finite number $k (\geq 4)$ of propositional symbols. Since the number of pairs (I, J) of sets of formulas constructed by those symbols are finite, providing every pair (I, J) satisfying $T_P(I) = J$ will produce those inference rules written as meta-rules. □

There are natural deduction systems that have more inference rules, while they are realized using those ten rules presented above. For instance, inference rules of contradiction: $p \wedge \neg p \vdash \bot$ is realized by MP by identifying $\neg p \equiv (p \rightarrow \bot)$; and $\bot \vdash q$ is realized by MP as $\bot \wedge (\bot \rightarrow q) \vdash q$. Thus, once learning the MP rule (13) by **LF1T**, we can get $\bot \vdash q$ or $p \wedge \neg p \vdash \bot$ by instantiating a propositional variable p or q with $\bot$.

Replacement rules for logically equivalent formulas, such as *De Morgan's law, communication, association, distribution, transposition* and *double negation*, are represented as $hold(F) \leftarrow hold(G)$ and $hold(G) \leftarrow hold(F)$ for $F \equiv G$ in background knowledge. Background knowledge is combined with inference rules induced by **LF1T** to produce new rules of inference. For instance, the replacement rule $p \rightarrow q \equiv \neg p \vee q$ is represented as $hold(p \rightarrow q) \leftarrow hold(\neg p \vee q)$ and $hold(\neg p \vee q) \leftarrow hold(p \rightarrow q)$ in background knowledge. Then together with the MP rule (13), we can get the new rule $hold(q) \leftarrow hold(\neg p \vee q), hold(p)$ (which is also an instance of **DS**).

3   Learning Non-logical Inference Rules

In Sect. 2 we consider a logical system $\mathcal{L}$ that has axiomatic systems for inference. $\mathcal{L}$ provides correct training data (S, T) to a machine learning algorithm $\mathcal{C}$ in Fig. 1, and **LF1T** successfully reproduces inference rules of $\mathcal{L}$ as a natural deduction system. On the other hand, axiomatic systems for inference do not always exist, especially for *non-logical* inference. In this case, a set of input-output pairs (or premise-consequence pairs) are not given from a teacher agent $\mathcal{A}$ in general, but can be implicitly hidden in log files of dynamic systems or in dialogues with unknown agents. A machine learning system $\mathcal{C}$ identifies those input-output relations automatically to produce a set of meta-theoretical inference rules for the domain or inference patterns of those agents. This section addresses three case studies for learning non-logical inference rules.

3.1   Abduction

A fallacy is an incorrect inference while often used in our daily life. For instance, the followings are well-known *logical fallacies* [19]:

Affirming a Disjunct: From $p \vee q$ and p, infer $\neg q$.
Affirming the Consequent: From $p \to q$ and q, infer p.
Denying the Antecedent: From $p \to q$ and $\neg p$, infer $\neg q$.

Affirming the consequent is also used for *abductive inference* [10]. In what follows, we address a case of learning inference rules for abduction.

Given background knowledge K and an observation $hold(G)$, abduction computes an explanation $hold(F)$ such that $hold(F \to G) \in K$. Unlike the case of learning deduction rules in Sect. 2.3, K, G and F are represented by propositional constants rather than propositional variables. This is because we assume no axiomatic system $\mathcal{L}$ in learning abduction rules. Training data are then provided as a particular background knowledge, an observation and its candidate explanations. The goal is to construct a general inference scheme of abduction from those data.

For example, let K be the background knowledge:

$$wet\text{-}grass \to wet\text{-}shoes$$
$$rained \to wet\text{-}grass$$
$$sprinkler\text{-}on \to wet\text{-}grass.$$

Using metalogic expression, the above rules are expressed as

$$K = \{\, hold(wg \to ws),\ hold(r \to wg),\ hold(s \to wg) \,\}$$

where ws, wg, r and s abbreviate *wet-shoes*, *wet-grass*, *rained* and *sprinkler-on*, respectively.

In computing rules for abduction, a state transition is given as a pair (I, J) such that I contains an observation O and rules R from background knowledge K, and J is an explanation such that $R \cup J \vdash O$. For instance, $(I, J) = (\{hold(wg \to ws),\ hold(ws)\}, \{hold(wg)\})$ means that the observation $hold(ws)$ is explained by $hold(wg)$ using $hold(wg \to ws)$ in K. To reduce the hypotheses space, we set *abducibles* as $\mathcal{H} = \{hold(wg), hold(r), hold(s)\}$ and assume any explanation as an element of $\mathcal{H}$. Let $\mathcal{B} = K \cup \{hold(wg), hold(ws), hold(r), hold(s)\}$. We apply the top-down **LF1T** for atoms in $\mathcal{H}$. Given pairs of transitions, **LF1T** finally produces the following rules:

$$hold(wg) \leftarrow hold(wg) \tag{14}$$
$$hold(wg) \leftarrow hold(wg \to ws), hold(ws) \tag{15}$$
$$hold(r) \leftarrow hold(r) \tag{16}$$
$$hold(r) \leftarrow hold(r \to wg), hold(wg) \tag{17}$$
$$hold(r) \leftarrow hold(r \to wg), hold(wg \to ws), hold(ws) \tag{18}$$
$$hold(s) \leftarrow hold(s) \tag{19}$$
$$hold(s) \leftarrow hold(s \to wg), hold(wg) \tag{20}$$
$$hold(s) \leftarrow hold(s \to wg), hold(wg \to ws), hold(ws) \tag{21}$$

Explanations are represented by atoms in the heads. The rule (14) means that the explanation wg self-explains the observation wg. The rule (15) means that the explanation wg is produced from the background knowledge $wg \rightarrow ws$ and the observation ws. Each rule represents abduction for different explanations. To obtain general inference rules for abduction, we use *least generalization* of Horn clauses [9,12]. In applying least generalization, we first classify rules into equivalent classes based on the number of atoms in the body of a rule. Given a rule R, $| b(R) |$ represents the number of atoms in the body of R. Then define the equivalence class as

$$C_i = \{ R : |b(R)| = i \}.$$

The rules (14)–(21) are classified into $C_1 = \{(14), (16), (19)\}$, $C_2 = \{(15), (17), (20)\}$, $C_3 = \{(18), (21)\}$. The rules in C_1 represent explanations as observations. The rules in C_2 represent explanations produced by one step of Affirming the Consequent. The rules in C_3 represent explanations produced by two steps of Affirming the Consequent. Abduction computes different explanations by $C_1 - C_3$, so state transitions (I, J) that are consistent with C_i are not always consistent with C_j $(j \neq i)$. For instance, the transition $(I, J) = (\{hold(r \rightarrow wg), hold(wg \rightarrow ws), hold(ws)\}, \{hold(r)\})$ is consistent with the rule (18), but is inconsistent with (15). So we compute rules $C_1 - C_3$ separately. The least generalization of C_2 is computed as follows. First, implication $p \rightarrow q$ is represented as a term $imp(p, q)$. Then the least generalization of (17) and (20) becomes

$$hold(F) \leftarrow hold(imp(F, wg)), hold(wg), hold(G)$$

where F and G are variables. After eliminating redundant atoms, it becomes[3]

$$hold(F) \leftarrow hold(imp(F, wg)), hold(wg) \tag{22}$$

which represents that any abducible F that implies wg is an explanation of the observation wg. Computing the least generalization of (22) and (15) and removing redundant atoms, we get

$$hold(F) \leftarrow hold(imp(F, G)), hold(G)$$

which represents that any abducible F that implies G is an explanation of the observation G. As such, computing least generalization of each set and removing redundant atoms, we obtain

$$hold(F) \leftarrow hold(F) \tag{23}$$
$$hold(F) \leftarrow hold(F \rightarrow G), hold(G) \tag{24}$$
$$hold(F) \leftarrow hold(F \rightarrow G), hold(G \rightarrow H), hold(H). \tag{25}$$

Each rule represents an inference rule of abduction. Among them, the rule (25) is obtained by a repeated application of (24), and the rule (23) is obtained by

---

[3] An atom A occurring in $b(R)$ is *redundant* if $b(R) \setminus \{A\} \equiv_\theta b(R)$ where $\equiv_\theta$ is equivalence under θ-subsumption $\leq_\theta$, i.e., $R_1 \leq_\theta R_2$ iff $h(R_1\theta) = h(R_2)$ and $b(R_1\theta) \subseteq b(R_2)$ for some substitution θ.

putting $F \equiv G$ in (24). As a result, we can pick the rule (24) as the inference rule of abduction.

The above example provides a simple case of learning abduction rules. On the other hand, there would be a case such that abduction is taken place together with deduction. For instance, suppose that a duplex system fails (sf) when two computers (c_1 and c_2) are down simultaneously. It is known that one of the computers does not work ($\neg c_1$) due to some trouble (tr). The situation is represented in the background knowledge as $K = \{\, tr,\ tr \to \neg c_1,\ \neg c_1 \wedge \neg c_2 \to sf \,\}$. Given the observation sf, a candidate explanation is $\neg c_2$, that is, c_2 does not work too. In this case, the next rule is produced

$$hold(\neg c_2) \leftarrow hold(tr), hold(tr \to \neg c_1), hold(\neg c_1 \wedge \neg c_2 \to sf), hold(sf)$$

representing that $\neg c_2$ is an explanation of the observation sf. By constructing a similar rule for $\neg c_1$, the next rule is produced by least generalization

$$hold(\neg F) \leftarrow hold(G), hold(G \to \neg H), hold(\neg F \wedge \neg H \to sf), hold(sf)$$

which means that if one of the two computers does not work by some reason, the system's failure is explained by the problem of another computer. The rule would be further generalized to

$$hold(F) \leftarrow hold(G), hold(G \to H), hold(F \wedge H \to K), hold(K) \qquad (26)$$

The rule (26) represents an inference where abduction is taken place with deduction. To verify the correctness of produced rules, they are applied to other test cases and check whether those rules provide appropriate explanations and accurate predictions.

3.2   Frame Axiom

Applying **LF1T** to state transitions describing the world change could enable us to learning *frame axioms* [7]. Let us consider a block world such that there are three blocks a, b and c where a is on b, b and c are on a table t. The state is represented as

$$S = \{hold(on(a,b)),\ hold(on(b,t)),\ hold(on(c,t)),\ hold(clear(a)),\ hold(clear(c))\}$$

where $clear(x)$ means that there is nothing on x. After moving a on top of c the state changes into

$$T = \{hold(on(a,c)),\ hold(on(b,t)),\ hold(on(c,t)),\ hold(clear(a)),\ hold(clear(b))\}.$$

Now we have only one state transition (S,T) then a state change from S to T is deterministically described using a *bottom-up* version of **LF1T** [5] that produces a transition rule from S to T. A bottom-up **LF1T** for definite logic programs is sketched below.

Bottom-up **LF1T**(E: pairs of Herbrand interpretations, P: a definite logic program)

1. If $E = \emptyset$ then output P and stop;
2. Pick $(I, J) \in E$, and put $E := E \setminus \{(I, J)\}$;
3. For each $a \in J$, let

$$R_a^I := a \leftarrow \bigwedge_{b_i \in I} b_i;$$

4. If R_a^I is not subsumed by any rule in P, then $P := P \cup \{R_a^I\}$ and simplify P by removing all rules subsumed by R_a^I;
5. Return to 1.

Given the state transition $(I, J) = (S \cup \{move(a, c)\}, T)$, the bottom-up **LF1T** produces

$$hold(on(a, c)) \leftarrow conj(S), move(a, c) \tag{27}$$
$$hold(on(b, t)) \leftarrow conj(S), move(a, c) \tag{28}$$
$$hold(on(c, t)) \leftarrow conj(S), move(a, c) \tag{29}$$
$$hold(clear(a)) \leftarrow conj(S), move(a, c) \tag{30}$$
$$hold(clear(b)) \leftarrow conj(S), move(a, c) \tag{31}$$

where $conj(S)$ is the conjunction of atoms in S. Among them, rules (28), (29), and (30) describe transitions that do not change by the action of $move(a, c)$, since $conj(S)$ contains the same atoms appearing in the head of each rule.

Suppose moving c on top of a at the state S. Using the state transition $(I, J) = (S \cup \{move(c, a)\}, T)$, the bottom-up **LF1T** produces

$$hold(on(c, a)) \leftarrow conj(S), move(c, a) \tag{32}$$
$$hold(on(a, b)) \leftarrow conj(S), move(c, a) \tag{33}$$
$$hold(on(b, t)) \leftarrow conj(S), move(c, a) \tag{34}$$
$$hold(clear(c)) \leftarrow conj(S), move(c, a) \tag{35}$$

Among them, rules (33), (34), and (35) describe transitions that do not change by the action of $move(c, a)$. Computing the least generalization of (30) and (35) and removing redundant atoms, we get

$$hold(clear(x)) \leftarrow conj(S), move(x, y) \tag{36}$$

The rule (36) represents a frame axiom saying that moving a block x to y at the state S does not change the clearness of x. Computing the least generalization of (28) and (34) and removing redundant atoms, we get

$$hold(on(b, t)) \leftarrow conj(S), move(x, y) \tag{37}$$

The rule (37) represents a frame axiom saying that moving a block x to y at the state S does not change the location of b on the table. Furthermore, generalizing (29) and (37), we get

$$hold(on(z,t)) \leftarrow conj(S), move(x,y) \qquad (38)$$

However, the rule (38) has the instance

$$hold(on(c,t)) \leftarrow conj(S), move(c,a)$$

which conflicts with the consequence of (32).[4]

Hence one can conclude that the rules (36) and (37) are valid frame axioms, while (38) is not a proper rule and is discarded. As such, frame axioms are successfully produced by induction.

3.3   Conversational Implicature

Non-logical inferences are also used in *pragmatics* [6]. In conversation or dialogue, the notion of *conversational implicature* [3] is known as a pragmatic inference to an implicit meaning of a sentence that is not actually uttered by a speaker. For instance, if a speaker utters the sentence "I have two children", it normally implicates "I do *not* have *more than* two children". This is called a *scalar implicature* which says that a speaker implicates the negation of a semantically stronger proposition than the one asserted. Given a collection of dialogues, a question is whether a machine learning system can automatically acquire pragmatic rules of inference that interpret implicit meaning behind utterance. To realize this, we assume a simple dialogue system that is able to converse with a human. Given a sentence S by a human, the system asks whether a sentence T that is semantically stronger than S is true or not. The human answers "yes" if it is true, and "no" otherwise. For instance, the following dialogue is taken place.

human: I have two children.
computer: Do you have three children?
human: No.

. . .

human: I have a car.
computer: Do you have two cars?
human: No.

The human's utterance is translated into factual knowledge as follows:

$$hold(have(child, s(s(0)))) \qquad (39)$$

$$hold(\neg have(child, s(s(s(0))))) \qquad (40)$$

$$hold(have(car, s(0))) \qquad (41)$$

$$hold(\neg have(car, s(s(0)))) \qquad (42)$$

---

[4] Here we assume the existence of a *state constraint*: $\forall x \forall y \, [hold(on(x,t)) \wedge hold(on(x,y)) \rightarrow y = t]$ asserting that if an object x is on a table t and x is on y then y is t.

Let (I, J) be a pair in which I represents the initial utterance of a human and J represents a reply by the human in response to a question by a computer. Then the above dialogue is represented by pairs:

$$(I, J) = (\{hold(have(child, s(s(0))))\}, \{hold(\neg have(child, s(s(s(0)))))\}),$$
$$(\{hold(have(car, s(0)))\}, \{hold(\neg have(car, s(s(0))))\}).$$

In this case, both I and J contain a single atom, so that the bottom-up **LF1T** simply constructs the transition rules as $A \leftarrow B$ where $A \in J$ and $B \in I$:

$$hold(\neg have(child, s(s(s(0))))) \leftarrow hold(have(child, s(s(0)))) \qquad (43)$$
$$hold(\neg have(car, s(s(0)))) \leftarrow hold(have(car, s(0))) \qquad (44)$$

The least generalization of (43) and (44) becomes

$$hold(\neg have(x, s(y))) \leftarrow hold(have(x, y)). \qquad (45)$$

The rule (45) means if one says that the number of x he/she has is y, it implies that the number is not $y + 1$. Suppose that the background knowledge contains the rule:

$$hold(\neg have(x, z)) \leftarrow hold(\neg have(x, w)), z \geq w \qquad (46)$$

which says that if one does not have x that is w in number, then he/she does not have x more than w. Then, after learning (45), the next rule is deduced by (45) and (46):

$$hold(\neg have(x, z)) \leftarrow hold(have(x, y)), z \geq s(y) \qquad (47)$$

which means that if one says that the number of x he/she has is y, it implies that he/she does not have it more than y. The rule (47) represents a rule of scalar implicature. On the other hand, if there is another dialogue such that

human: I have two dollars.
computer: Do you have three dollars?
human: Yes.

Then the next rule is constructed

$$hold(have(dollar, s(s(s(0))))) \leftarrow hold(have(dollar, s(s(0)))) \qquad (48)$$

This is an exceptional case of scalar implicature. In the presence of (48), the rule (47) could be refined as

$$hold(\neg have(x, z)) \leftarrow hold(have(x, y)), z \geq s(y), x \neq dollar$$

or one could construct scalar implicature rules for individuals such that:

$$hold(\neg have(child, z)) \leftarrow hold(have(child, y)), z \geq s(y) \quad \text{(from (43) and (46))}$$
$$hold(\neg have(car, z)) \leftarrow hold(have(car, y)), z \geq s(y) \quad \text{(from (44) and (46))}$$

As such, scalar implicature rules are inductively constructed.

4   Discussion

New paradigms are emerging in machine learning such as *ontology learning* [18] and *representation learning* [1]. Also recent advances in robotics argue possibilities of robots' recognizing objects in the world, categorizing concepts, and associating names to them (*physical symbol grounding*) [2]. Once robots successfully learn concepts and associate symbols to them, the next step is to learn relations between concepts and logical or physical rules governing the world. According to Piaget's theory of cognitive development, children start learning concepts and symbols at age earlier than two (*pre-operational stage*), and begin to understand logical or rational thought at age around seven (*concrete operational stage*) [11]. Representation learning and physical symbol grounding aim at realizing machine learning at the level of the pre-operational stage. On the other hand, learning inference considered in this paper targets the problem of realizing machine learning at the level of the concrete operational stage.

In learning logical inferences, we show that the **LF1T** induction algorithm can reconstruct a system $\mathcal{L}$ of natural deduction. An interesting question is whether a machine learning algorithm can discover a *new* axiomatic system that is semantically equivalent to $\mathcal{L}$. It addresses the possibility of AI's discovering new logics that are unknown to human mathematicians. A logical formulation of conversational implicature is studied in [15] while, to the best of our knowledge, learning conversational implicature from dialogue has never been explored. We showed a simple case study of learning scalar implicature in conversation. If one develops an AI that learns and understands conversational implicature, it will realize an intelligent chat bot that can understand implicit meaning of humans' utterance. Learning non-logical inference also involves the problem of learning thinking patterns of individuals. It would be interesting to investigate whether a system can learn thinking patterns of people in particular regions or in particular professionals.

This paper realizes learning inference as induction of meta-rules. Induction in meta-theories are proposed in [4,8]. Inoue *et al.* [4] introduce *meta-level abduction* to invent predicates and apply it to finding physical skills. Muggleton *et al.* [8] introduce *meta-interpretive learning* to invent relations by abduction and apply it to learning grammatical rules. These studies represent background knowledge as a meta-theory and abduce rules as meta-facts, while their goals are not learning inference rules. Induction or ILP has mostly been used for learning knowledge, while little study has been devoted to the topic of learning inference. The current study argues the possibility of using ILP for learning inference and serves as a step for opening the topic.

5   Conclusion

This paper studied learning inference by induction. We first addressed a method of learning deductive inference rules using the **LF1T** induction algorithm. We showed that **LF1T** successfully produces inference rules of propositional natural deduction as transition rules from premises to consequences. Secondly, we applied the method to learning non-logical inference rules. We showed that abductive inference rules are obtained from observations and explanations, frame axioms are computed by state changes, and scalar implicature rules in conversation could be learned from simple dialogues.

This is a preliminary research for learning inference rules by induction, and the proposed method will need further elaboration and extension in practice. Providing all possible transitions, **LF1T** will output production rules that are minimal with respect to subsumption. A limitation is that the number of possible transactions increases exponentially in proportion to the size of the Herbrand base. Further optimization is needed for learning inference rules from huge data. In this paper, **LF1T** is used for learning propositional inference rules in Sect. 2. For learning first-order inference rules, provide pairs of premises-consequences at the fact level and produce ground rules at first, then generalize those rules using ILP technique such as least generalization. Such a technique is directly applied to quantifier-free rules and further technique will be needed for learning quantified formulas. Applying the proposed framework to learning inference rules in other logics (e.g. probability/fuzzy logic), learning social rules in multiagent systems (e.g. negotiation), and learning strategic rules in games (e.g. chess) would be of interest.

References

1. Bengio, Y., Courville, A., Vincent, P.: Representation learning: a review and new perspectives. IEEE Trans. Pattern Anal. Mach. Intell. **35**, 1798–1828 (2013)
2. Coradeschi, S., Loutfi, A., Wrede, B.: A short review of symbol grounding in robotic and intelligent systems. KI - Kunstliche Intelligenz **27**, 129–136 (2013)
3. Grice, H.P.: Logic and conversation. In: Cole, P., Morgan, J. (eds.) Syntax and Semantics, 3: Speech Acts, pp. 41–58. Academic Press (1975)
4. Inoue, K., Furukawa, K., Kobayashi, I., Nabeshima, H.: Discovering rules by meta-level abduction. In: De Raedt, L. (ed.) ILP 2009. LNCS, vol. 5989, pp. 49–64. Springer, Heidelberg (2010)
5. Inoue, K., Ribeiro, T., Sakama, C.: Learning from interpretation transition. Mach. Learn. **94**, 51–79 (2014)
6. Levinson, S.C.: Pragmatics. Cambridge University Press, Cambridge (1983)
7. McCarthy, J., Hayes, P.J.: Some philosophical problems from the standpoint of artificial intelligence. In: Meltzer, B., Michie, D. (eds.) Machine Intelligence, vol. 4, pp. 463–502. Edinburgh University Press, Edinburgh (1969)
8. Muggleton, S.H., Lin, D., Pahlavi, N., Tamaddoni-Nezhad, A.: Meta-interpretive learning: application to grammatical inference. Mach. Learn. **94**, 25–49 (2014)
9. Nienhuys-Cheng, S.H., de Wolf, R.: Foundations of Inductive Logic Programming. LNCS (LNAI), vol. 1228. Springer, Heidelberg (1997)
10. Peirce, C.S.: Collected papers of Charles Sanders Peirce. In: Hartshorne, C., Weiss, P., Burks, A. W. (eds.) Harvard University Press (1958)
11. Piaget, J.: Main Trends in Psychology. Allen & Unwin, London (1973)
12. Plotkin, G.D.: A note on inductive generalization. In: Meltzer, B., Michie, D. (eds.) Machine Intelligence 5, pp. 153–63. Edinburgh University Press (1970)

13. Prawitz, D.: Natural Deduction: A Proof-Theoretical Study. Dover Publications, Mineola (2006)

14. Ribeiro, T., Inoue, K.: Learning prime implicant conditions from interpretation transition. In: Davis, J., et al. (eds.) ILP 2014. LNCS, vol. 9046, pp. 108–125. Springer, Heidelberg (2015). doi:10.1007/978-3-319-23708-4_8

15. Sakama, C., Inoue, K.: Abduction and conversational implicature. 12th International Symposium on Logical Formalizations of Commonsense Reasoning, AAAI Spring Symposium, Technical report SS-15-04, pp. 130–133 (2015)

16. Sakama, C., Inoue, K.: Can machines learn logics? In: Bieger, J., Goertzel, B., Potapov, A. (eds.) AGI 2015. LNCS, vol. 9205, pp. 341–351. Springer, Heidelberg (2015)

17. van Emden, M.H., Kowalski, R.A.: The semantics of predicate logic as a programming language. J. ACM **23**, 733–742 (1976)

18. Wong, W., Liu, W., Bennamoun, M.: Ontology learning from text: a look back and into the future. ACM Comput. Surv. **44**, 20:1–20:36 (2012)

19. Woods, J., Irvine, A., Walton, D.: Argument: Critical Thinking, Logic and the Fallacies. Prentice-Hall, Toronto (2000)

Identification of Transition Models of Biological Systems in the Presence of Transition Noise

Ashwin Srinivasan[1], Michael Bain[2(✉)], Deepika Vatsa[3], and Sumeet Agarwal[3]

[1] Birla Institute of Technology and Science Pilani, Goa Campus, Pilani, India
ashwin@goa.bits-pilani.ac.in
[2] School of Computer Science and Engineering, University of New South Wales,
Sydney, Australia
m.bain@unsw.edu.au
[3] Department of Electrical Engineering, Indian Institute of Technology Delhi,
New Delhi, India
{eez138262,sumeet}@iitd.ac.in

Abstract. The identification of transition models of biological systems (Petri net models, for example) in noisy environments has not been examined to any significant extent, although they have been used to model the ideal behaviour of metabolic, signalling and genetic networks. Progress has been made in identifying such models from sequences of qualitative states of the system; and, more recently, with additional logical constraints as background knowledge. Both forms of model identification assume the data are correct, which is often unrealistic since biological systems are inherently stochastic. In this paper, we model the transition noise that can affect model identification as a Markov process where the corresponding transition functions are assumed to be known. We investigate, in the presence of this transition noise, the identification of transitions in a target model. The experiments are re-constructions of known networks from simulated data with varying amounts of transition-noise added. In each case, the target model traces a specific trajectory through the state-space. Model structures that explain the noisy state-sequences are obtained based on recent work which formulates the identification of transition models as logical consequence-finding. With noisy data, we need to extend this formulation by allowing the abduction of new transitions. The resulting structures may be both incorrect and incomplete with respect to the target model. We quantify the ability to identify the transitions in the target model, using probability estimates computed from transition-sequences using PRISM. Empirical results suggest that we are able to identify correctly the transitions in the target model with transition noise levels ranging from low to high values.

1 Introduction

Chemical equations are symbolic statements not of what will happen, but of what may happen. Thus, the equation $2H_2(g) + O_2(g) \rightarrow 2H_2O(g)$ does not mean that hydrogen and oxygen will necessarily react to produce water (the "g"

© Springer International Publishing Switzerland 2016
K. Inoue et al. (Eds.): ILP 2015, LNAI 9575, pp. 200–214, 2016.
DOI: 10.1007/978-3-319-40566-7_14

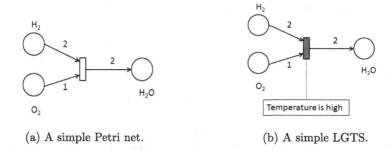

(a) A simple Petri net.   (b) A simple LGTS.

Fig. 1. Two transition system representations of the reaction $2H_2 + O_2 \rightarrow 2H_2O$.

denotes the reactants and products are in a gaseous state). Filling a balloon with the reactants, for example, does not immediately result in balloon full of water vapour. Additional conditions may be needed (in this case, a high temperature) for the reaction to occur. Even if external conditions are favourable, it is possible that the reaction may not proceed. There is thus a non-determinism associated with any chemical processes, including those that occur in cells [6].

Our interest in this paper is in computational models of biological networks, like Petri nets, in which processes are *transitions* representing local changes to the qualitative state of the system. An example of a Petri net representation of the "water" reaction is in Fig. 1a.

We extend our previous work to study identification of transition models of biological systems in the presence of added transition noise. We develop a two-stage method (Fig. 2). First, deductive and abductive inference is used in a logic programming framework [17] to identify all transitions consistent with observational data. Second, probabilistic logic programming [15] is used to estimate parameters for the identified transition system. Evaluation is by reconstruction experiments on benchmark biological systems with varying levels of added transition noise. In Sect. 2 we describe our approach. Empirical results and discussion are in Sect. 3. We discuss some related work in Sect. 4, and conclude in Sect. 5.

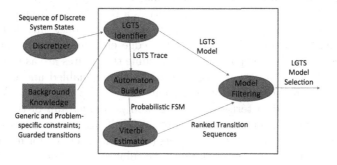

Fig. 2. System identification with noisy data.

2 Transition Identification Under Transition Noise

Biochemical Background. A theory of reactions based on collisions requires not only that the reactant molecules collide with each other, but that they should collide in correct orientations. The reaction:

$$A - B + C \rightarrow C - A + B$$

will only occur if C collides with the complex $A - B$ on the "A-side". The probability of such collisions can be mathematically modelled [8]. With small numbers of $A - B$ and C, this reaction may simply not occur. Such non-determinism can be due to extrinsic and intrinsic effects, which has been studied in the chemical and biological literature [6]. In particular, sources of biological noise are both intrinsic, due to the inherent stochasticity of processes of the system such as gene expression, and extrinsic, due to conditions in the environment. In this paper we focus on intrinsic noise.

Petri Nets. A Petri net [3] is a bipartite directed graph with two finite sets of nodes, called places and transitions. Arcs are either from a place to a transition, so the place is an input for the transition, or from a transition to a place, in which case the place is an output of the transition. Transitions have a finite number of input places and a finite number of output places. Places can be occupied by zero or more tokens, and arcs can be labelled with an integer weight greater than or equal to one. In a Petri net, a transition is enabled and can hence be executed, or "fired", when the number of tokens at each input place is equal to or greater than the weight on the corresponding arc.

LGTS. Transition systems [10] are a formalism used to model the behaviour of dynamic systems. In this paper we adopt the framework of Logical Guarded Transition Systems (LGTS) [16,17], a generalisation of Petri nets [3] based on logic programming. For example, Fig. 1b shows the "water" Petri net as an LGTS, with the constraint that temperature should be high for the reaction to proceed, which would be encoded as a logic program. Essentially, an LGTS extends the Petri net formalism by allowing logical *guarded* transitions, specifying the pre, post and invariant conditions that must hold for the transition to occur. An LGTS for some system comprises a relation *lgts* from sequences of observed system states to sequences of guarded transitions, plus definitions of guarded transitions, and background knowledge encoding domain-specific and generic constraints. In this paper we will assume that states are as in Petri nets, i.e., vectors of place-values [3]. For transitions to be enabled and executed we additionally require the *guard* of transitions to evaluate to $TRUE$. A guard of a transition is defined in terms of the *pre-* and *post-condition* of the transition and it's *invariant*. The guard evaluates to TRUE if the pre- and post-condition and the invariant are all $TRUE$. Petri nets (and their extended form, that allows "read" and "inhibit" arcs, and variations like coloured and stochastic Petri nets) can be seen as special cases of LGTS's. LGTS's can be used to simulate system behaviour: given some initial state, output a sequence of transitions and their corresponding states. LGTS's can also be identified from samples of system behaviour: given a sequence of states, determine a corresponding sequence of transitions.

Transition Identification. The identification of LGTS's can be formulated as logical consequence finding [17]. The basic system identification task is as follows:

> **Given:** (a) A sequence S of states, representing observations of the system behaviour; and
> (b) Background knowledge B containing generic and domain-specific constraints and definitions of guarded transitions; and
> (c) The definition of a relation $G = lgts(S, T)$ that is TRUE for all pairs S and T s.t. T is an LGTS model of S.
>
> **Find:** An acceptable LGTS. Any LGTS T satisfying $B \land G \Rightarrow \exists T \ lgts(S, T)$ is acceptable.

If B and G can be encoded as logic programs, then the T's can be computed using the usual theorem prover used by logic programming systems. With a bound on the number of tokens allowed in each place, the LGTS models for an observation sequence S can be computed by a non-deterministic finite automaton (NFA) [17]. The NFA is a transducer that reads zero or one input symbols (observations) and writes out the corresponding transition.

Transition Noise. We can distinguish three kinds of "noise" that can affect the identification of transition-based models from data: (a) signal noise (for example, there is an error in the concentration of a metabolite); (b) state noise (for example, a gene is incorrectly recorded as being "on" when it is actually "off"); and (c) transition noise (for example, a reactant is not produced when it normally should be). In principle, both (b) and (c) can be modelled by a Markov process, if the corresponding transition functions were known, which is the assumption made in this paper.

Specifically, we model *transition noise* as a probabilistic transition system. A transition system [10] is a pair $(S, \rightarrow)$ where S is set of states and $\rightarrow$ is a binary relation on S modelling the set of transitions. In a *deterministic* transition system, $\rightarrow$ is a function, i.e., for a given state s there is at most one successor state s' such that $s \rightarrow s'$. In a *non-deterministic* transition system states may have more than one successor state, i.e., $\rightarrow: S \times \mathcal{P}(S)$. To model *probabilistic* transitions a probability distribution is defined on each non-deterministic transition.

Identification in the Presence of Transition Noise. Viewing an LGTS as an NFA is useful when considering identification from noisy data. The presence of noise means that there is some mismatch between transitions in background knowledge and the observations. This incompleteness has two aspects: first, there may be some missing intermediate states; second, states may have incorrect values with respect to the corresponding transitions. The first problem is solved based on [4] by generating values for missing place vectors during LGTS identification; in terms of NFA execution this is equivalent to allowing the "empty" input ϵ to correspond to an output transition. The second is solved by allowing the abduction of transitions for "noisy" state pairs; this is akin to employing a theorem-prover that uses SOLD-resolution [18].

These identification steps for noisy observational data will typically lead to an expanded set of transitions. To determine which transitions are more likely to model the underlying system, and which are simply due to noise, the final step in our approach is to sample repeatedly from the distribution over noisy transitions and construct a probabilistic automaton. The resulting probabilities can be used to rank output LGTS models for user inspection, and to evaluate the method.

3   Empirical Evaluation

Our goal is to investigate the identification of the transitions comprising a system, given noisy data sequences representing system behaviour. Specifically: we intend to investigate if the transitions involved in generating the ideal sequence of states can be identified given sufficient numbers of noisy data sequences.

3.1   Problems

The investigation considered system identification for the following problems, listing in order of increasing model-size:

Water. The well-known school-level problem of the formation of water from hydrogen and oxygen forms the simplest system we will examine. The problem clearly consists of a single reaction involving 3 kinds of molecules (places).

MAPK. The MAPK pathway is a protein-based sequence of events that translate a signal at the cell-surface to the nucleus. The pathway commences when a protein or a hormone binds to a receptor protein that is usually bound to the cell-membrane. This triggers a sequence of events that stops with the DNA expressing one or more genes that alter cell function. At any one step of the cascade, phosphor groups are attached to proteins. This phosphorylated form of the protein then forms a "switch" for commencing the next step. MAPK is a central signalling pathway that is used in all cell-tissues to communicate extra-cellular events to the cell nucleus. It is used to regulate a variety of responses, like hormone action, cell-cycle progression and

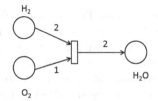

Fig. 3. Network model for the formation of water.

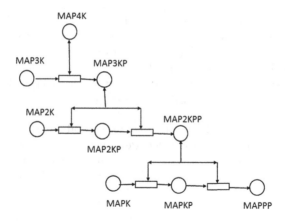

Fig. 4. Network model of the MAPK cascade.

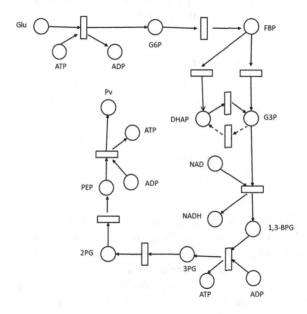

Fig. 5. Network model of the glycolysis pathway. The conversion of DHAP to G3P is taken to be in one-direction only (the reverse is shown by a dashed line, and not identified).

cell-differentiation. It is also of immense clinical value, since a defect in the pathway often leads to uncontrolled growth. Proteins in the pathway are thus natural targets for anti-cancer drugs.

Glycolysis. The glycolysis pathway was the first metabolic pathway to be discovered. It is a classic case of a series of metabolic reactions in which products of one reaction form the substrates (reactants) for the next reaction. The glycolysis pathway is comprised of 10 such reactions. The reactions breakdown (metabolize) each molecule of glucose into two molecules of pyruvate. The sequence proceeds in three stages: primary (3 reactions), splitting (2 reactions) and phosphorylation (5 reactions). Altogether, 15 metabolites are involved. The pathway is one of the central metabolic pathways in living organisms: it provides an essential part of the energy required for the functioning of a cell, and is used in several metabolic processes.

We note that although Glycolysis contains more metabolites and reactions than MAPK, the latter requires an extended Petri net [3] for its representation (it requires "read" arcs), while Glycolysis (and Water) can be represented by normal Petri nets.

3.2   Data

In this paper, data will be taken to be the result of one or more experiments, each resulting in a table of the following form:

Places	States					
	s_0	s_1	s_2	$\ldots$	$\ldots$	s_k
p_1	$s_{1,0}$	$s_{1,1}$	$s_{1,2}$	$\ldots$	$\ldots$	$s_{1,k}$
p_2	$s_{2,0}$	$s_{2,1}$	$s_{2,2}$	$\ldots$	$\ldots$	$s_{2,k}$
$\ldots$	$\ldots$	$\ldots$	$\ldots$	$\ldots$	$\ldots$	$\ldots$
$\ldots$	$\ldots$	$\ldots$	$\ldots$	$\ldots$	$\ldots$	$\ldots$
p_l	$s_{l,0}$	$s_{l,1}$	$s_{l,2}$	$\ldots$	$\ldots$	$s_{l,k}$

Places are, as noted above, as in the literature on Petri nets. In experiments in this paper they are restricted to Boolean values (with 0 denoting that the quantity represented by a place is absent, and 1 denoting that it is present in a sufficient quantity). Each table (experimental result) of this form gives rise to a sequence $(s_0, s_1, \ldots, s_n)$. A noisy data sequence will contain a sequence of states that will differ from an ideal (noise-free) sequence of states.

System behaviours are thus sequences of system states of the form $S_i = (s_{i,0}, s_{i,1}, \ldots, s_{i,n_i})$. Each such sequence can be taken as a set of state-pairs $\{(s_{i,0}, s_{i,1}), (s_{i,1}, s_{i,2}), \ldots, (s_{i,n_i-1}, s_{i,n_i})\}$; and a set of sequences $S = \{S_1, S_2, \ldots, S_j\}$ can be represented by the union of the corresponding sets of state-pairs. We will call this set $StatePairs(S)$.

3.3   Models

An LGTS trace for a state-pair (s_i, s_f) is a set $Trace(s_i, s_f) = \{T_1, T_2, \ldots, T_k\}$, where $T_1 = (t_1, r_1, m_0, m_1), T_2 = (t_2, r_2, m_1, m_2), \ldots, T_k = (t_k, r_k, m_{k-1}, m_k)$, where: (a) each t_j is a guarded transition; (b) $r_j = m_j - m_{j-1}$; and (c) $s_i = m_0$; and (d) $s_f = m_k$. That is, $m_1, m_2, \ldots m_{k-1}$ are intermediate states.

An LGTS model for a state-pair (s_i, s_f) is $T(s_i, s_f) = \{(t, r) : (t, r, m_a, m_b) \in Trace(s_i, s_f)\}$. It is straightforward to extend this to a set of sequences and state-pairs. Given a set of sequences $S = \{S_1, S_2, \ldots, S_j\}$, let $TracePairs(S) = \bigcup_{(s_i, s_j) \in StatePairs(S)} Trace(s_i, s_j)$. Then $LGTS(S) = \{(t, r) : (t, r, m_a, m_b) \in TracePairs(S)\}$.

It is possible to construct a non-deterministic finite-state automaton (NFA) from $TracePairs(S)$ that can output all the sequences in S. Further, a probabilistic finite-state automaton (PFA) can be constructed from the NFA by extending the transition function to include a probability distribution. We omit proofs of these claims here, and show some example automata instead (see Figs. 8 and 9).

3.4   Algorithms and Machines

Simulated data and probability estimates of the sequences of transitions are obtained using the probabilistic environment provided within the PRISM system [15]. LGTS models are obtained using a Prolog program that implements the basic system identification task in Sect. 2. All programs were run on a Lenovo dual processor Core i7 laptop, running a Linux emulation with 4 GB of memory.

3.5   Method

Our method is straightforward:

1. Repeat R times:

   For low, medium and high noise levels:
   i. Generate N noisy data sequences, using probabilistic versions of transitions in the ideal model
   ii. Obtain LGTS proofs for each (noisy) data sequence
   iii. Using the transition sequences in the LGTS proofs, determine the extent to which the correct transitions can be identified. This may involve an abduction step, somewhat related to SOLD-resolution [18].

The following details are relevant:

1. Low, medium and high levels of noise are defined in terms of the probability with which a probabilistic transition generates the output-state of the corresponding deterministic transition. Here, low noise means that this probability is 90 %; medium noise means that the probability is 75 %; and high noise means that the probability is 50 %.
2. In this paper, $R = 10$. We generate $N = 100$ noisy data sequences for each model. The data are generated using a Hidden Markov (HMM) model, in which states are observed and transitions are hidden.[1] Mainly, two kinds of

---

[1] This may be slightly confusing in the first instance, since in HMMs, states are hidden. Here we are referring to biological system states and biological system transitions.

probabilities have to be specified for the HMM: the conditional probability of emitting a (biological system) state, given a (biological system) transition; and the conditional probability of a transition, given a transition. One additional probability distribution is needed that allows an initial transition to be selected randomly. With each deterministic transition in the model t with input state s_i and output state s_j we associate a probabilistic variant with input s_i and a set S_j as output, with $s_j \in S_j$. The probability with which the HMM emits elements of S_j is determined by the noise level (thus, with low noise, the probability that s_j is emitted is 0.9 and so on). Here, the set S_j consists of s_j and all states within a 1-bit Hamming distance of s_j. The transitions that can follow t with output state s_j are all transitions that have s_j as an input state. The HMM selects amongst these uniformly. This simulation is done using the PRISM program.

3. The efficacy of transition identification is computed as follows. Let the target model of the system consist of a set of transitions T_{act}. Let the system model consist of the set of transitions T_{pred}. We only seek Viterbi probabilities of such sequences of length $|T_{act}|$ (this is provided as a length-bound on the sequences considered by PRISM). The NFA has no loops, and thus $|T_{pred}| = |T_{act}|$. For each experimental run, we compute $E = |T_{act} - T_{pred}|/|T_{act}|$, which is, in effect a false-negative rate. Since the sets T_{act} and T_{pred} are of the same size, the false-positive rate is equal to the false-negative rate. $E = 0$ denotes perfect identification and $E = 1$ denotes perfect mis-identification.

3.6   Results

Table 1 shows some supplementary details related to the transition-identification problem.

Some of the success see in the results of Table 3 may be attributed to the amount of data provided: the results in Table 3 are from 100 (simulated) sequences of observed values. In practice, each such sequence can be thought of as an experiment; and 100 experiments is unusual in Biology, unless dealing with some form of high-throughput automation like microarray data generation.

Table 1. Number of transitions (T) in the logical LGTS model and the number of transitions (T^*) identified using the probabilistic model (the latter includes an initial dummy transition). The quantities in parentheses are the standard deviations obtained from multiple repeats of the identification task. All numbers are rounded up to the nearest integer, since fractional transitions are meaningless.

Noise	Water		MAPK		Glycolysis	
	$|T|$	$|T^*|$	$|T|$	$|T^*|$	$|T|$	$|T^*|$
Low	16 (1)	2 (0)	101 (12)	6 (0)	255 (48)	11 (0)
Med	19 (2)	2 (0)	160 (8)	6 (0)	425 (23)	11 (0)
High	22 (1)	2 (0)	245 (10)	6 (0)	625 (75)	11 (0)

Table 2. Identification of transitions with small datasets. Here 10 observation sequences are used to identify the transitions.

Noise	Error E		
	Water	MAPK	Glyc.
Low	0.00 (0.00)	0.00 (0.00)	0.00 (0.00)
Med	0.00 (0.00)	0.03 (0.10)	0.20 (0.10)
High	0.55 (0.35)	0.88 (0.20)	0.88 (0.10)

Table 3. Identification of transitions in the target model. E denotes the average false-negative rate (that is, the fraction of true transitions not identified). Both the predicted and actual transition sequences are of the same length (see text for details), so the false-negative and false-positive rates are the same. Thus a value of 0.0 for E denotes all–and only–the transitions in the model are identified; and a value of 1.0 denotes none of the transitions in the model are identified. The column P denotes the average Viterbi probability of the highest ranking transition sequence (transitions in this sequence are predicted as being in the target model). Noise levels of low, medium, and high refers to transitions having a probability of 10 %, 25 % and 50 % of resulting in an incorrect state. The quantities in parentheses are the standard deviations obtained from multiple repeats of the identification task.

Noise	Error E			Probability P		
	Water	MAPK	Glyc.	Water	MAPK	Glyc.
Low	0.00 (0.00)	0.00 (0.00)	0.00 (0.00)	0.81 (0.05)	0.55 (0.03)	0.31 (0.04)
Med	0.00 (0.00)	0.00 (0.0)	0.00 (0.00)	0.55 (0.05)	0.17 (0.04)	0.04 (0.01)
High	0.00 (0.00)	0.20 (0.00)	0.90 (0.10)	0.23 (0.04)	0.01 (0.00)	0.01 (0.00)

Table 2 shows the results obtained with significantly fewer observation sequences (10, instead of 100). Now, identification becomes more difficult at moderate noise-levels.

These results suggest that when small amounts of data are the norm, we can expect probabilistic transition-identification to work best at low levels of transition-noise. While this is perhaps obvious enough, a caveat is nevertheless worth noting. A good case can be made that noise-levels that we have labelled here as "medium" and "high" are unlikely to be encountered in practice (for example, a chemical reaction is very unlikely to result in unexpected products 50 % of the time). We would therefore expect the transition-identification approach proposed here to work well in practice, even if the data instances are few in number.

Results related to identifying transitions in the target models are in Table 3. The results show clearly that identification of the set of target transitions is perfect at low and medium noise-levels, and only degrades when transition noise is high. It is also apparent that performance is degraded by higher levels of noise, evident from the decrease in Viterbi probability. It is surprising that identification performance is so good, because although the target models studied are relatively simple, with Water, MAPK and Glycolysis having 1, 5 and 10 transitions with 3, 9 and 15 places, respectively, the hypothesis space of *possible* transitions is quite large in each case.

3.7   Transition Identification Worked Example: Water

Shown in Figs. 6, 7, 8 and 9 are examples of the main stages in our approach for the "Water" transition system. In Fig. 6 we see three sample observational state sequences. These are value vectors for the named places over the sequence. In Fig. 7 we show inferred *transitions* in LGTS trace pairs for these sequences, comprising the transition name, difference (reaction) vector, and the corresponding predecessor and successor states for these transitions. Figures 8 and 9 show the constructed NFA and PFA, respectively, for this system.

Data S:

$$([(h2, 0), (o2, 0), (h2o, 0)], [(h2, 1), (o2, 1), (h2o, 0)], [(h2, 0), (o2, 0), (h2o, 1)])$$
$$([(h2, 0), (o2, 0), (h2o, 0)], [(h2, 1), (o2, 1), (h2o, 0)], [(h2, 0), (o2, 0), (h2o, 1)])$$
$$([(h2, 0), (o2, 0), (h2o, 0)], [(h2, 1), (o2, 1), (h2o, 0)], [(h2, 1), (o2, 0), (h2o, 1)])$$

$\cdots$

$\cdots$

Fig. 6. Transition identification worked example: Noisy data sequences simulating formation of water from hydrogen and oxygen; states shown as place-value tuples.

Transition $TracePairs(S)$:

$$(t1, [(h2, 1), (o2, 1), (h2o, 0)], [(h2, 0), (o2, 0), (h2o, 0)], [(h2, 1), (o2, 1), (h2o, 0)])$$
$$(t2, [(h2, -1), (o2, -1), (h2o, 1)], [(h2, 1), (o2, 1), (h2o, 0)], [(h2, 0), (o2, 0), (h2o, 1)])$$

$\cdots$

$\cdots$

Fig. 7. Transition identification worked example: LGTS trace for data of Fig. 6.

The transitions in the highest ranked transition sequence from the PFA (ranked by Viterbi probability obtained using PRISM built-in predicates) are used to identify the system transitions. In Fig. 9, the highest ranked sequence is $(t1, t2)$, and the system model is taken to be $\{(t1, r1), (t2, r2) : (t1, r1), (t2, r2) \in LGTS(S)\}$. Note that this will usually be a subset of $LGTS(S)$ and in some sense, can be considered a generalisation of that set.

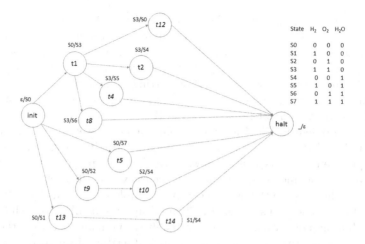

Fig. 8. Transition identification worked example: Non-deterministic finite-state automaton obtained from LGTS trace that correctly derives all data. Transition-identifiers in italics are result of abduction steps made by the theorem-prover to obtain the LGTS trace. Sx/Sy denotes input/output symbols of transition. All states are connected to "halt" state" — most omitted for clarity.

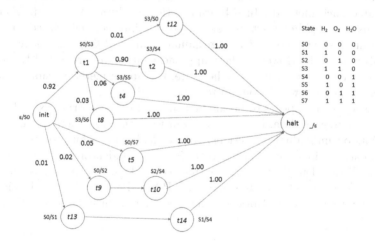

Fig. 9. Transition identification worked example: Probabilistic finite-state automaton obtained from the NFA in (c). Probabilities are estimated using transition sequences followed by the NFA when emitting data sequences as output.

4   Related Work

There have been several approaches to the identification of Petri nets from data. For example, in the area of business process mining (BPM) [1] a number of methods construct Petri nets. An important problem in BPM is process discovery, i.e., given a log of activity sequences from some software system, generate a model of the business processes involved. This model is often a Petri net, generated, for example, by first extracting a transition system from the activity log and then synthesising a Petri net from it [11]. Although this method is well-founded with theoretical guarantees [5] it has exponential worst-case time complexity and in practice it can lead to both under- and overfitting, so most implementations use additional heuristics [1]. Nonetheless, it has a natural declarative formulation [11] so could provide a basis for further work in ILP.

Another problem in BPM is that of conformance checking, which has been investigated in ILP [12] by supervised learning of integrity constraints to detect non-compliance in activity logs. This was extended to a probabilistic framework using Markov logic and shown to increase accuracy and efficiency [2]. However, these approaches require the use of negative example traces, which are typically not available in biological settings. In [7] an unsupervised ILP approach to BPM was used, but did not use a formalised probabilistic model. In one sense the above BPM tasks are easier than the standard biological setting, since log files typically contain many repeated instances of activity sequences, thus providing enough data to disambiguate Petri net models. In biological settings this is usually not the case due to the difficulty of running many repeated biological experiments.

One heuristic approach often adopted in BPM has been genetic algorithms, and a number of authors working on Petri net identification in biology have also investigated such methods. In [14] grammatical evolution was used, where the hypothesis space of Petri net models for genetic interactions related to disease was specified using a context-free grammar, and in [13] the incidence matrix was evolved directly. However, these approaches were only reported to work on networks of up to four genes. Furthermore, the above methods cannot identify extended Petri nets, which contain read and inhibitory arcs [17].

The work of [4] appears to be the first method to reconstruct extended Petri nets from time series data. In contrast to our work this approach does not enable constraints on individual transitions. Although there is a form of abduction in this approach, it does not allow for *inductive* steps in identification, as in [16,17], or probabilistic identification, as in this paper. Learning from interpretation transitions was presented in [9] where an ILP-based approach is used to identify Boolean network models. However, this does not identify probabilistic transitions.

5   Conclusion

We have studied the identification of transition models of biological systems under conditions of added transition noise, extending our previous work. Using the probabilistic logic programming system PRISM [15] we have modelled varying levels of transition noise in three benchmark biological systems. We apply a two-step method, first using a logic-programming approach incorporating both deduction and abduction to identify a complete set of logical guarded transitions that explain the noisy state-sequences. This logical model is then used to construct a probabilistic finite-state automaton. The parameter-estimates obtained (using PRISM) for this automaton are used to identify a subset of the logical model that is then taken to be the system model. Our experimental results show that the method can reconstruct known networks from simulated data with varying amounts of transition-noise.

There are some immediate ways in which the empirical evaluation could be extended. First, we have presented some evidence that as data size decreases, system identification is affected at high noise levels. However, "smaller" data sizes here is still much higher than what is normally available in experimental life-sciences. For example, how will system identification be affected when data are available from 2 to 3 experiments? Secondly, we have seen that larger sized networks are affected more by high noise-levels than networks of smaller size. Although the high noise-levels used here are unlikely to be encountered in real data, there is nevertheless a need for further work to investigate the effect of network size. It is possible that, even at low noise levels, system identification may degrade for very large networks. Third, although we have examined the effect of incorrect data, we have not examined the effect of incomplete data. The probabilistic setting we have used naturally accounts for this by the use of the EM algorithm to estimate probabilities. We should therefore be able to investigate the effect of missing data on system identification.

References

1. Van Der Aalst, W.: Process mining: overview and opportunities. ACM Trans. Manage. Inf. Syst. **3**(2), 7 (2012)
2. Bellodi, E., Riguzzi, F., Lamma, E.: Probabilistic declarative process mining. In: Bi, Y., Williams, M.-A. (eds.) KSEM 2010. LNCS, vol. 6291, pp. 292–303. Springer, Heidelberg (2010)
3. David, R., Alla, H.: Discrete, Continuous, and Hybrid Petri Nets, 2nd edn. Springer, Berlin (2010)
4. Durzinsky, M., Wagler, A., Marwan, W.: Reconstruction of extended Petri nets from time series data and its application to signal transduction and to gene regulatory networks. BMC Syst. Biol. **5**, 113 (2011)
5. Ehrenfeucht, A., Rozenberg, G.: Partial (Set) 2-Structures: Part II: state spaces of concurrent systems. Acta informatica **27**, 343–368 (1990)
6. Elowitz, M., Levine, A., Siggia, E., Swain, P.: Stochastic gene expression in a single cell. Science **297**, 1183–1186 (2002)
7. Ferilli, S., Esposito, F.: A logic framework for incremental learning of process models. Fundamenta Informaticae **128**, 1–31 (2013)
8. Gillespie, D.: Exact stochastic simulation of coupled chemical reactions. J. Phys. Chem. **81**(25), 2340–2361 (1977)
9. Inoue, K., Ribeiro, T., Sakama, C.: Learning from interpretation transition. Mach. Learn. **94**(1), 51–79 (2014)
10. Keller, R.: Formal verification of parallel programs. Commun. ACM **19**(7), 371–384 (1976)
11. Kindler, E., Rubin, V., Schäfer, W.: Process mining and petri net synthesis. In: Eder, J., Dustdar, S. (eds.) BPM Workshops 2006. LNCS, vol. 4103, pp. 105–116. Springer, Heidelberg (2006)

12. Lamma, E., Mello, P., Riguzzi, F., Storari, S.: Applying inductive logic programming to process mining. In: Blockeel, H., Ramon, J., Shavlik, J., Tadepalli, P. (eds.) ILP 2007. LNCS (LNAI), vol. 4894, pp. 132–146. Springer, Heidelberg (2008)
13. Mayo, M.: Learning petri net models of non-linear gene interactions. Biosystems **82**(1), 74–82 (2005)
14. Moore, J., Boczko, E., Summar, M.: Connecting the dots between genes, biochemistry, and disease susceptibility: systems biology modeling in human genetics. Mol. Genet. Metab. **84**, 104–111 (2005)
15. Sato, T., Kameya, Y.: PRISM: A symbolic-statistical modeling language. In: Proceedings of the 15th International Joint Conference on Artificial Intelligence (IJCAI 1997), pp. 1330–1335 (1997)
16. Srinivasan, A., Bain, M.: Knowledge-guided identification of petri net models of large biological systems. In: Muggleton, S.H., Tamaddoni-Nezhad, A., Lisi, F.A. (eds.) ILP 2011. LNCS, vol. 7207, pp. 317–331. Springer, Heidelberg (2012)
17. Srinivasan, A., Bain, M.: Identification of Transition-Based Models of Biological Systems using Logic Programming. Technical report UNSW-CSE-TR-201425, School of Computer Science and Engineering, University of New South Wales, Sydney, Australia, December 2014
18. Yamamoto, A.: Representing inductive inference with SOLD-resolution. In: Proceedings of the IJCAI 1997 Workshop on Abduction and Induction in AI (1997)

Author Index

Printed in the United States
By Bookmasters